SCIENTIFIC WRITING

A READER AND WRITER'S GUIDE

JEAN-LUC LEBRUN

Trainer of researchers and scientists from A*STAR Research Institutes, Singapore
Former Director, Apple-ISS Research Centre, Singapore

SCIENTIFIC WRITING

A READER AND WRITER'S GUIDE

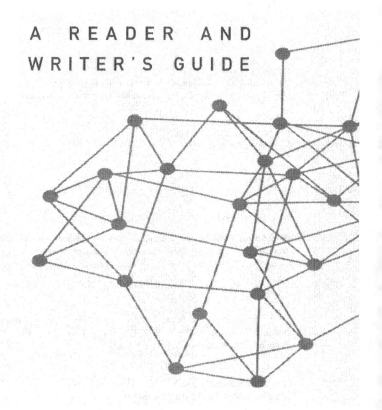

 **World Scientific**

W JERSEY • LONDON • SINGAPORE • BEIJING • SHANGHAI • HONG KONG • TAIPEI • CHENNAI

Published by

World Scientific Publishing Co. Pte. Ltd.

5 Toh Tuck Link, Singapore 596224

USA office: 27 Warren Street, Suite 401-402, Hackensack, NJ 07601

UK office: 57 Shelton Street, Covent Garden, London WC2H 9HE

Library of Congress Cataloging-in-Publication Data
Lebrun, Jean-Luc.
    Scientific writing : a reader and writer's guide / Jean-Luc Lebrun.
    p. cm.
    Includes index.
    ISBN-13 978-981-270-473-3
    ISBN-10 981-270-473-6
    ISBN-13 978-981-270-144-2 (pbk)
    ISBN-10 981-270-144-3 (pbk)
    1. Technical writing.   I. Title.
    T11.L455 2007
    808.0665--dc22

                                      200829578

British Library Cataloguing-in-Publication Data
A catalogue record for this book is available from the British Library.

First published 2007
Reprinted 2008, 2009

Typeset by Stallion Press
Email: enquiries@stallionpress.com

Printed by FuIsland Offset Printing (S) Pte Ltd. Singapore

Preface

You know how to write grammatically correct English. Congratulations. You have read Strunk and White's little book, *The Elements of Style*.[a] Perfect. To pursue true writing excellence, you now need to take into consideration the people key to your success: the readers.

What readers fear the most while reading a scientific paper is to get stuck or left behind. They are stuck when the experienced writer zigzags around the familiar obstacles in the knowledge field, whilst readers crash into them; and they are left behind when the knowledgeable writer runs where they can only walk. The knowledge gap that separates you from your readers cannot be ignored, yet adequate background knowledge does not guarantee that motivated readers will find reading your paper easy and rewarding. Much more is required of them. A scientific paper requires more memory, attention, and time than a typical novel of the same length. Good writing should therefore take into account the reader's ignorance, fatigue, short-term memory, and impatience in order to minimise their impact.

Unique writing techniques rarely presented in books on technical writing will bring the writer closer to the six qualities that are the hallmark of great scientific writing: fluid, organised, clear, concise, convincing, and interesting (FOCI). Consider sentence structure. Does

---

[a] Strunk W Jr and White EB, *The Elements of Style*, Penguin Press, New York, 2005.

placing a conjunction such as *because, but,* or *although* at the head of a sentence provide more reading pull than placing it midway? Consider sentence progression. Does keeping the topic of the first sentence constant throughout a paragraph help the reader progress smoothly through a written argument? Consider the reader's expectations. Can a single word in a sentence trigger large expectations? "Because it was raining that day," creates the expectation that the writer will explain what happened because of the rain. The sentence finishes with "the paint did not dry on time." The reader reaches the end of the sentence knowing why the paint did not dry-the first expectation raised is fulfilled, but another expectation arises: the paint did not dry on time for what? Expectations drive reading forward in science as they do in literature. By creating and controlling pull, progression, and expectations, the writer can guide the reader.

Readers have different expectations for each part of a scientific article, from its title to its conclusions. Since ignoring these expectations frustrates readers, the writer should avoid the short introduction that sheds little light on the "what" and "why" of the paper, the abstract that is indistinguishable from the conclusions, the misleading title, the baggy structure, and the immature and unprocessed visuals. This book will help writers learn how to put together a coherent set of parts that satisfies readers.

This book comes with a metaphorical box of chocolates: 48 stories designed to liven up reading and reinforce the learning process. It also comes with a core of 100 examples inspired or quoted from scientific articles. No attempt has been made to "sweeten" them. Do not let them intimidate you. What is of importance in each of these examples is not their impact on the world of science: it is the placement of the words in the sentence and the expectations they create.

This book was written at the request of many scientists who have participated in the scientific writing skills seminars I conduct in various parts of the globe. In their assessment of the course, the participants highlighted benefits; some expected, some unexpected. As expected, those who had already published papers felt that their writing had improved by keeping the reader in mind. Junior scientists without any publishing experience were relieved that they no longer had to blindly imitate the work of others, not knowing whether what they were imitating was good or bad. Unexpectedly, even senior scientists with great publishing experience found that the seminar had improved their analytical reading skills and had equipped them with a method to conduct better peer reviews.

Before turning the page, words of appreciation are due. More than 1000 scientists from many research centres helped me to understand and love the scientific reader. This book is dedicated to them. Three authors, through their books, influenced the contents of this book: Michael Alley[b] on scientific writing, George Gopen[c] on reader energy and expectations, and Don Norman[d] on user interfaces. They have my deepest respect. They are the giants on whose shoulders I climbed to discover a new world they had explored well before I did. If, thanks to them, I discovered new techniques that will be of help to the reader of this book, may they share the credit.

The expanded edition is making atonement for the scientists who found in these pages recommendations to break the canonical rules of the systematic use of the past tense and passive voice. It also features a new chapter expanding the horizon of those eager to go beyond the teachings of this book. Over seven years of constant research on the best practices in scientific writing, I have bookmarked websites that contain excellent articles worth reading and even printing as reference. This chapter contains my bookmarks and a brief description of what you will find on the web pages.

---

[b] Alley M, *The Craft of Scientific Writing*, Springer, New York, 1997.
[c] Gopen GD, *Expectations: Teaching Writing from the Reader's Perspective*, Pearson Longman, 2004.
[d] Norman D, *The Design of Everyday Things*, Basic Books, New York, 2002.

Contents

Part I

The Reading Toolkit

This title probably conjures up the image of a schoolboy's pencilcase containing a few chosen articles designed to help reading: a pair of glasses, a bookmark, instant coffee, etc. However, this toolkit is quite special. It contains resources invisible to the naked eye, like time, memory, energy, attention, and motivation. A skillful writer minimises the time, memory, and energy needed for reading, while keeping reader attention and motivation high.

1

Require Less from Memory

The Forgotten Acronym

Let us start with a story.

> **A reading accident**
>
> Peter reads an article from the proceedings of a conference. He follows the text in a linear fashion. Suddenly, he stops, places his index finger underneath a word, and rapidly scans the text he has just read, searching for something. What he is looking for is not on the page. With his left hand, he flips back one page, and then another ... he stops again. His face lights up. Satisfied, Peter flips back to the page he was reading before this unexpected and unwelcomed reading U-turn, and sets his eyes back to where the index finger marks the place for reading to resume. What happened? A reading accident: the forgotten acronym. Peter probably encountered an unfamiliar acronym defined only once by the author at the beginning of the paper. Peter had read its definition, but time had passed and he had forgotten it.

Acronyms allow writing to be more concise. However, conciseness is unhelpful if it decreases clarity. An acronym is clear within the paragraph in which it is defined. If it continues to be used regularly in the paragraphs that follow, the reader is able to keep its meaning in mind. But, if it appears irregularly or if reading is frequently interrupted, the acronym — away from the warm nest of the reader's short-term memory — loses its meaning. Food gets cold fast when it is out of the oven; you warm it up before eating it. Similarly, keep the acronym warm in the reader's memory; redefine it regularly in your paper.

Reader curiosity or impatience also contributes to reading accidents. The reader has the unfortunate habit (from the author's point of view) of skipping entire parts of your article to go directly to a figure, or to a section in your paper that seems interesting (via the heading or subheading). If the figure caption or the heading/subheading contains acronyms, and if the reader has skipped the sections that contain their definitions, then the accident will happen.

Avoiding problems with acronyms is easy:

- If an acronym is used only two or three times in the entire paper, it is better not to use one at all (unless it is as well known as IBM).
- If an acronym is used more than two or three times, expand its letters the first time it appears on a page so that the reader does not need to flip pages back and forth. Some journals ask authors to regroup all acronyms and their definitions at the beginning of their paper so that the reader can locate them more easily.
- Avoid acronyms in visuals or define them in their caption.
- Avoid acronyms in headings and subheadings because readers often read the structure of a paper before going inside the paper.
- Be conservative. Define all acronyms, except those commonly understood by the readers of the journal where your paper is published.

> ### The Singapore taxi driver
>
> The other day, while I was in Singapore, I hailed a taxi. I wanted to go to a research institute located on the campus of Nanyang Technological University (NTU). The taxi stopped. I got in and said, "Nanyang Technological University, please." The taxi driver, an old man who had clearly been doing this job for many years, replied, "I do not know where it is." His answer surprised me. The university is old and well established; surely he had taken passengers there before. I started explaining that it was at the end of the expressway towards Jurong . . . all of a sudden, his face lit up and he said with a large smile, "Ah! You mean NTU!" That day, I learned that an acronym is sometimes better known than its definition.

Notice the just-in-time definition of the acronym in the following example.

The new universal learning algorithm SVM (support vector machine) had a profound impact on the world of classification.

The Detached Pronoun

This, it, them, they, and *their*[a] are all pronouns. A pronoun usually replaces a noun, but sometimes it replaces a phrase, a sentence, or even a full paragraph. Like the acronym, it is a shortcut that avoids the repetition of words.

Pronouns and acronyms are both pointers. This characteristic is at the root of all problems:

1. If you point in the direction of someone who has already left the room, nobody will understand. Likewise, if the noun the

---

[a] "Their" is not technically a pronoun: it is a possessive pronominal adjective, but it functions as a pronoun. In the French language, "their" (*leur*) is a demonstrative pronoun.

pronoun points to is 20 or 30 words back in the text, it may have left the reader's short-term memory; the noun–pronoun link is broken. Usually, this memory lapse is not enough to discourage readers from reading forward. They tolerate ambiguity and read on because they are hopeful that the text will become clearer later. Interpretation errors and reduced understanding are therefore likely.

2. If you point towards a person in a group far away from you, people will find it difficult to guess whom exactly you are pointing to. When the pronoun points back to several likely candidates, the reader — whose incomplete understanding of the text does not allow disambiguation — will pick the most likely candidate and read on, hoping clarity will be forthcoming. If that likely candidate is the wrong one, then interpretation errors will follow and understanding will drop to a lower level.

3. Finally, some fingers seem to point nowhere; actually, they point somewhere, but only the person who is pointing knows where. When the pronoun points to something that is only in the mind of the author, the reader is left guessing and more often than not guesses wrongly. Understanding thus drops to a lower level.

A diagram (☛1) helps to visualise the exploratory process followed by readers when encountering a pronoun.

> The new notation ☛1 is simply an invitation to look at visual **1**.
>
> I do not mention whether visual **1** is a diagram, a table, or a photo because you know the difference.
>
> Why the big black ☛ before the number? It is to help your eyes easily return to the text at the right place after you have looked at the visual. As you return, just let your eyes be guided by the dark beacon.

The diagram highlights that a reader stops searching for another candidate (i.e. antecedent) **as soon as** a likely one is found in his or her

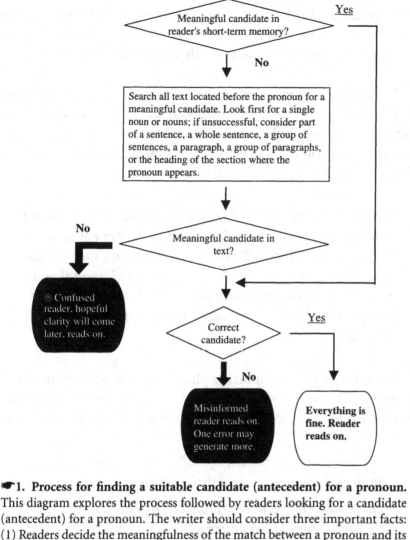

☛ 1. Process for finding a suitable candidate (antecedent) for a pronoun.
This diagram explores the process followed by readers looking for a candidate
(antecedent) for a pronoun. The writer should consider three important facts:
(1) Readers decide the meaningfulness of the match between a pronoun and its
candidate. Therefore, their knowledge of the topic is part of the process, and
little knowledge may mean greater ambiguity. (2) Readers stop looking for a
candidate as soon as they think they have found one (preferably in their short-
term memory). They do not have the energy, the time, or the will to stop and
analyse whether the pronoun candidate they chose is the correct one. As a result,
errors occur more frequently if the candidate is so distant from its pronoun
that it no longer is in memory. (3) Readers continue reading whether they
have identified the correct candidate or not. Being unable to find the candidate
may be less "damaging" to the understanding than continuing reading with a
"corrupted" understanding because, in the first case, the reader seeks to increase
understanding, while in the second case, the reader is lulled into a false sense of
understanding.

short-term (working) memory. The choice of candidate is influenced by the reader's knowledge: the more superficial the knowledge, the more error-prone the choice will be. Authors who wish nonexperts to read their paper should be aware that pronouns present dangers.

In the following example, try and determine what the pronoun "*their*" refers to. The three candidates are in bold. Had the sentence been clear, this task would have been instantaneous. You will probably struggle; but if you do not, ask yourself how much does knowledge of the field assist you in making the correct choice.

> *The cellular automaton (CA) cell, a natural candidate to model the electrical activity of a cell, is an ideal component to use in the simulation of **intercellular communications**, such as those occurring between cardiac cells, and to model **abnormal asynchronous propagations**, such as **ectopic beats**, initiated and propagated cell-to-cell, regardless of the complexity of THEIR patterns.*

It is difficult to determine the plural noun pointed to by "*their*" because the sentence segment "*regardless of the complexity of their patterns*" could be moved around in the sentence and still make sense.

> *... to use in the simulation of intercellular communications, regardless of the complexity of their patterns ...*
>
> *... to model abnormal asynchronous propagations, regardless of the complexity of their patterns ...*
>
> *... such as ectopic beats, regardless of the complexity of their patterns ...*

Communications, propagations, and beats can all display complex patterns. Let us decide that in this text, "*their*" represents the "*abnormal asynchronous propagations.*"

The ambiguity can be removed in different ways. First, one could simply omit the detail if it is not essential, or make that point later in the paragraph. The long sentence would then be seven words shorter.

*The cellular automaton (CA) cell, a natural candidate to model the electrical activity of a cell, is an ideal component to use in the simulation of **intercellular communications**, such as those occurring between cardiac cells, and to model **abnormal asynchronous propagations**, such as **ectopic beats**, initiated and propagated cell-to-cell.*

One could also rewrite the sentence to make the pronoun disappear.

The cellular automaton (CA) cell — a natural candidate to model the electrical activity of a cell — is an ideal component to use in the simulation of intercellular communications, such as those occurring between cardiac cells, and to model the cell-to-cell initiation and propagation of abnormal asynchronous events (such as ectopic beats) with or without complex patterns.

Finally, one could repeat the noun instead of using a pronoun.

The cellular automaton (CA) cell, a natural candidate to model the electrical activity of a cell, is an ideal component to use in the simulation of intercellular communications, such as those occurring between cardiac cells, and to model abnormal asynchronous events, such as ectopic beats, initiated and propagated cell-to-cell, however complex the propagation pattern may be.

In science, clarity overrides elegance; therefore, repeat to avoid ambiguity.

Search for the following words in your paper: *this, it, they, their,* and *them.*

If you were the reader, could you easily identify what the pronoun refers to without ambiguity? If you could not, remove the pronoun and repeat the noun(s)/phrase it replaces. An alternate route consists in rewriting the whole sentence in a way that removes the need for the pronoun.

The Diverting Synonym

Bis repetitas placent

That day, I could not understand why the paragraph I was reading was so obscure. The usual culprits were absent: the grammar was correct and the sentence length was average for a scientific article. I had noticed that words were repeated, but repetition usually clarifies and does not blight understanding. I decided to try and remove some of the repeated words. I then discovered the problem: four synonymous expressions.

1. *Known or predefined location.*
2. *Predefined location information.*
3. *Preprogrammed location information.*
4. *Identifiable position information.*

The author could have continued the game and added a few more synonymous expressions:

5. *Identifiable location information.*
6. *Predefined position information.*
7. *Preprogrammed position information.*

After removal of the synonyms, the structural problems appeared clearly. The paragraph was thus easier to rewrite.

Your language teacher may have told you to avoid repeating nouns within a sentence or in consecutive sentences. The advice given was, "Use synonyms, demonstrate your knowledge of the vast English vocabulary." In science, however, synonyms confuse readers, particularly those not familiar with the specialised terms used in your field. Therefore, avoid synonyms. Make your writing clear by consistently using the same keywords, even if it means repeating them. As an added benefit, you will lessen the demands on the memory of your readers: fewer new words also means less to remember.

The Distant Background

> ### The Macintosh factory
>
> When I moved to Cupertino, California, in 1986 to work at the headquarters of Apple Computer, I visited their Macintosh factory in Fremont. Every day, truckloads of components and parts came in, just enough for one day's production; and every day, containers of Macintoshes were shipped out. The net result: no local storage, no warehousing. I was witnessing a very efficient technique: just-in-time (JIT) manufacturing.

Traditionally, the background material the reader needs to understand your contribution is written in the first part of your article. If this background material is not used immediately, the memory will have to store it for later use. Unfortunately, the memory warehouse is small and the warehouse keeper is quite busy.

> ### The variable types
>
> There are two types of variables in a computer programme: global and local variables. Global variables
>
> *(Continued)*

> *(Continued)*
>
> are declared at the beginning of a programme and are known throughout the programme. Local variables are known only within the subroutine where they are declared. This interesting concept allows the computer to manage its memory space more efficiently. Global variables require permanent storage, whereas local variables free up their temporary memory storage space as soon as the programme exits the subroutine. Could this wonderful concept apply to writing?

Parking all background material in the introductory sections of your paper increases the demands on the reader's memory. Background material comes in two forms: the global background, applicable to the whole paper; and the local or just-in-time background, useful only to one section or paragraph of your paper. The just-in-time background imposes no memory load: it immediately precedes or follows what it makes clear. Here is a just-in-time example:

> *Additional information is readily available from "context" — other words found in the vicinity of the word considered.*

In this example, the word *"context"* is defined as soon as it appears.

When a heading or subheading in your paper contains a word requiring an explanation, explain it in the first sentence under the heading, in a just-in-time fashion.

Lysozyme solution preparation
Lysozyme, an enzyme contained in egg white, . . .

In this subheading, the word *"lysozyme"* is unusual. The writer defines it in the first sentence of the section.

The English language offers many ways to add just-in-time information. The "*lysozyme*" example uses an apposition — an expression that clarifies what comes before it. Kept short, appositions are very effective. Kept long, they are ineffective, as the following sentence demonstrates.

> *Lysozyme, a substance capable of dissolving certain bacteria, and present for example in egg white and saliva but also tears where it breaks down the cell wall of germs, is used without purification.*

Appositions are also ineffective when they slow down reading, which happens quite often when many are found midsentence.

> *The cellular automaton (CA) cell,* **a natural candidate to model the electrical activity of a cell,** *is an ideal component to use in the simulation of intercellular communications,* **such as those occurring between cardiac cells,** *and to model the abnormal asynchronous propagations,* **such as ectopic beats,** *initiated and propagated cell-to-cell, regardless of the complexity of their patterns.* [1 sentence, 57 words]

The sentence above is long because it is attempting to describe two things at one time. Reading would be faster if the sentence was divided into two homogeneous parts:

> *The cellular automaton (CA) cell is used in the simulation of intercellular communications because it can model the complex evolution of cell-initiated and cell-propagated signals in time and space. CA is therefore used here to model the electrical signals of cardiac cells, including those leading to abnormal asynchronous propagations such as ectopic beats.* [2 sentences, 54 words]

The Broken Couple

The hot tap

Do you remember the last time you stood still, hands under the hot water tap, waiting for the water to become warm, wasting cold water down the sink? Felt frustrated? When reading a sentence in which the verb never seems to arrive, has it occurred to you that your reader may also "waste" or ignore the words that separate the subject from its verb?

Details inserted between the main components of a sentence burden (burden comes from the old French *bourdon*, a "hum or buzz" — but do we need to know that!) the memory because they move apart two words that the reader expects to see together, such as the verb ("*burden*") and its object ("*the memory*") in this sentence. Such details

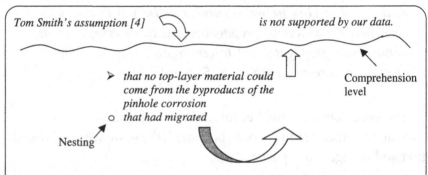

☛2. Sinking below the comprehension level. The nesting of subordinates has the same effect as plunging the reader below the comprehension level. In the end, what will count and be remembered is above the comprehension level, and what will be discarded as detail and forgotten is below the comprehension level. Two causes lead to the progressive confusion of the reader: (1) the phrase "*the byproducts of the pinhole corrosion*" that creates distance between the relative pronoun "*that*" and its antecedent "*byproducts*". It is not the corrosion that migrates, but rather the byproducts; and (2) the two nested subordinates starting with "*that*". To avoid the nesting, the writer could have changed the second subordinate into a noun, as in "*that no top-layer material could come from the migration of the pinhole corrosion byproducts*".

are often wasted, like cold water from a hot water tap. Separating the subject and the verb, as illustrated in ☞2, can be devastating.

Another couple of neighbours are best kept close: the visual and its full explanation. We no longer live in the days of silent movies. A visual must "tell all" by itself, without the need for text outside of its caption. Unless visuals are self-contained, the reader has to constantly shuttle back and forth between text and visual. Therefore, explain visuals **fully** in their caption.

You would do well to use the just-in-time principle and keep the following couples happily wedded:

• An unfamiliar word & its definition • An acronym & its definition • A noun/phrase & its pronoun • A verb & its subject	• A verb & its object • Background information & the text it clarifies • A visual & its complete caption

In summary, acronyms, pronouns, abusive detailing, background "ghettos", cryptic captions, and separated phrases all take their toll on the reader's memory.

Memory registers

I remember studying the structure of the Intel 8085 microprocessor back in 1981 (carbon-14 could not tell my age any better). I discovered that rapid access to memory is so critical to the overall speed of a micro-processor that the central processing unit (CPU) has its own dedicated memory registers right on the chip, or under the same roof, so to speak. Storing and retrieving data from these internal registers is ultrafast compared
(Continued)

(Continued)
to the time spent retrieving data from external memory. Like the CPU, do keep syntactically or semantically closely related items on the same page, in the same paragraph, in the same sentence, or on the same line. The reader will appreciate the increase in reading speed and the ease of understanding.

The Word Overflow

Our working memory is very similar to the rewriteable electronic memory. To be retained in memory, the information needs to be rewritten a number of times (it is therefore a slower process than the reading process). Furthermore, the current used to "imprint" the memory is greater than the current used to read its contents. The current, in the reader's case, is attention. It takes a great deal of attention. The process is also slow. Have you ever been able to absorb complex road directions without asking the person to repeat them? Going too fast creates an overflow. Working memory is not very elastic; it can be overstretched by a sudden word overflow.

"The main difference between the new micro molding machine design and the conventional 'macro' molding machines with reciprocating screw injection system is that by separating melt plastication and melt injection, a small injection plunger a few millimetres in diameter can be used for melt injection to control metering accuracy, and at the same time a screw design that has sufficient channel depth to properly handle standard plastic pellets and yet provide required screw strength can be employed in micro molding machines."[b]

---

[b] Zhao J, Mayes RH, Chen GE, Xie H, and Chan PS, "Effects of process parameters on the micro molding process", *Polymer Eng Sci* 43(9):1542–1554, 2003. © 2003 Society of Plastics Engineers.

This last sentence has a grand total of 81 words! Its syntax is acceptable and the meaning is clear enough for a specialist familiar with the machine, but the working memory necessary to process it is too large for most readers. Restructuring the sentence, breaking it down into logical segments, helps to reduce the demands on the working memory.

In conventional "macro" molding machines with reciprocating screw injection, melt plastication and melt injection are combined within the screw-barrel system. In the new micro molding machine, screw and injector are separated. The redesigned screw still has enough strength and channel depth to handle standard plastic pellets; but the separate injection plunger, now only a few millimetres in diameter, can be used to control the metering accuracy.

The rewritten paragraph has three sentences instead of one, and 66 words instead of 81. As a result, because our memory can handle it, clarity is increased.

In conclusion, if you want the reader to sail through your paper with minimal memory load, identify and remove the causes for overload.

 **Read your introduction again. Can you push background details closer to what they really explain? Are the sentences that feel long also ambiguous? Are details keeping apart elements of a sentence that should be closer?**

2

Sustain Attention to Ensure Continuous Reading

> *(Continued)*
>
> remained silent long enough for her to know that there could not have been many. "I can't think of one!" he said finally, "Even my own papers bore me." She asked, "Don't they teach you how to write interesting papers at university, you know, papers that attract the attention of scientists?" Vladimir sighed. "Attracting is fine," he said, "It is sustaining the attention which is hard. I wish I knew of some good attention-getters to keep my readers as awake and interested as you are."

Drama and suspense naturally seem out of reach for the scientific writer. However, besides scientific achievement, the writer is not as deprived of attention-getters as one might think. To capture attention, rely on five principles: move ideas forward, make important things stand out, illustrate to clarify, question to engage, and recreate suspense.

Move Ideas Forward

Change, in all its forms, is a great attention-getter. Take a change in paragraph, for example. The reader expects the story to progress, widen, narrow, or jump. The absence of change has the opposite effect. Sometimes the author stops ideas from moving forward. Puddles of details stagnate here and there, unconnected to the main stream. Sometimes the author, caught in a whirlpool, goes around in circles, repeating things that are already clear to the reader. When ideas are not in motion, two things happen to a paragraph: its length grows and its cohesion decreases. Additional length is often a consequence of paraphrasing. Needless paraphrases slow down reading and reduce conciseness.

When ideas are not in motion, two things happen to a paragraph: its length grows and its cohesion decreases. Additional *length is often a consequence of paraphrasing. **With paraphrasing, the paragraph lengthens without actually moving the ideas forward, since the sentences have the same meaning.** Needless paraphrases slow down reading and reduce conciseness.*

(Sentence 3, in bold, repeats what sentences 1 and 2 already cover.)

Additional length also occurs when details explain details. Nested detailing diverts attention. It takes the reader away from the main intent of the paragraph. The next paragraph is about the process of embryonic cell proliferation in a culture dish. The reader is distracted by an in-depth description of the culture dish (dish → coating → reason for coating), which could have been described in an earlier paragraph.

*For the next 3 days, the 30 embryonic cells proliferate in the culture dish. **The dish, made of plastic, has its inner surface coated with mouse cells that, through treatment, have lost the ability to divide, but not their ability to provide nutrients. The reason for such a special coating is to provide an adhesive surface for the embryonic cells.** After proliferation, the embryonic cells are collected and put into new culture dishes, a process called "replating". After 180 such replatings, millions of normal and still undifferentiated embryonic cells are available. They are then frozen and stored.*

The reader is distracted when the author returns to a point several sentences after the point is made just to add detail, as in the next example. In this case, sentence 4 (the last one) should immediately follow sentence 1.

After conducting microbiological studies on the cockroaches collected in the university dormitories, we found that their guts carried staphylococcus, members of the coliform bacilli, and other dangerous microorganisms when outside of the intestinal tract. Since they regurgitate food, their vomitus contaminates their *body. Therefore, the same microbes, plus moulds and yeasts, are found on the surface of their hairy legs, antennae, and wings. **It, is not astonishing to find such microorganisms in their guts, as they are also present in the human and animal faeces on which they feed.***

To reduce excessive paragraph length, follow these three steps: keep the main supportive details that contribute the most to your argument, and trim the rest; join and consolidate related details that are scattered; and restructure the paragraph to remove repetition and inconsistent keywords.

Sometimes, additional length is caused by lack of focus. The paragraph accumulates points and issues that are interwoven and difficult to disentangle without a complete restructure of the long paragraph.

Look at your long paragraphs and ask yourself, what am I trying to achieve with this paragraph? Does that support my overall contribution? What issue am I presenting or what point am I making? Is it the first time I am making this point? Can I make that point using fewer arguments, fewer words, or a figure? Am I making more than one point? Am I covering more than one issue? Would making two paragraphs out of this one paragraph clarify things and keep ideas in motion?

Make Important Things Stand Out

Subheadings attract attention because they stand out. Therefore, make your subheadings as informative and indicative of content as possible. Avoid hollow pointers such as "Simulation" or "Experiment". Some headings, however, are naturally hollow, such as "Introduction", "Discussion", or "Conclusions". They reveal the function, not the contents, of a section. They are standard and allow rapid navigation to parts of interest to the reader.

Within a paragraph, it is also possible to make things stand out. Take the **change in sentence length**, for example. After a long sentence, and particularly at the end of a paragraph, a short sentence carries much emphasis, as you will see. Why? Its syntax is usually simple and fast to process. Because it does not contain many words and is less technical, it is easier to understand. The last sentence of the following paragraph is four times shorter than the longest sentence that precedes it. Indeed, the pace quickens as the paragraph unfolds its sentences: 21 words, 27 words, 22 words, 17 words, and 6 words.

Photo annotation, a tedious manual task, is a labour of love towards future generations or a nostalgic revisiting of the past. For paper photos in albums or shoeboxes, annotations are either implicit (event-, time-, or subject-based) or explicit (scribbles underneath or on the back of a photo). For digital photos, annotations like time, date, and sometimes location (GPS coordinates) are automatically embedded in the file format by the camera. Could major life events (e.g. birthdays, weddings) or familiar scenery (e.g. beaches, mountains) also be automatically annotated? For a given culture, they can.

Underlining a sentence attracts attention. Underlining is one of many **changes in format and style** that act as eye magnets. Used in moderation, a numbered list, a box around text, **bold**, <u>underlined</u>, or italic text, a change in font, etc. are equivalent to raising the volume of your voice, or changing its pitch or inflexion. They break the monotony of paragraphs and make things stand out (note that the publisher may limit your choices by imposing a standard format).

Repetition is another effective way to tell the reader what you consider most important (the reader may not know without your help). Often used in conversations, repetition is not welcomed in writing, where it is a sign of an immature paper. However, there are two situations where it is deliberate and useful: to restate and rephrase your contribution, and to provide a summary at the end of a particularly difficult or long section.

Most writers say that a contribution is repeated four times in a paper: in the abstract, the introduction, the results section, and the conclusions. Some say five because they include the title. In fact, there are seven opportunities to strengthen your contribution through repetition: title, abstract, introduction, the body of text, conclusions, visuals, and subheadings. This repetition is not achieved through "copy-and-paste" or through a paraphrase using synonyms. It is a thoughtful re-presentation (presenting anew) of the contribution at varying levels of detail, using different tenses (more on that in part II of the book).

The summary, another repetition, clarifies what is important by rephrasing the section's main points succinctly and differently. It also gives readers a second chance to understand, and gives writers the assurance that readers will be able to keep in step with them.

> Words such as *to summarise, in summary, in other words, see Fig. X, in conclusion, in short, and briefly put* all perk up the attention of readers. They sustain interest and announce consolidation of knowledge.

Words conveying importance guide attention. They act like pointing fingers and are quite effective if used sparingly.

> Words such as *more importantly, significantly, notably, in particular, particularly, especially, even, and nevertheless* all help the reader to focus on what you consider important.

Illustrate to Clarify

Reading is hard, but writing is harder. Distilling years of research in less than 10 pages is a dangerous exercise. Like compressed audio files, compressed knowledge loses clarity. Even if the structure of your paper is clear, you need to reintroduce detail into your text to keep things clear.

The need for **examples** is not just a byproduct of the distillation process. Illustrative details are needed because, more often than not, your readers are not familiar with what is happening in your field of research. They may be scientists in the same domain (not field), but the distance between you and them in terms of knowledge is great, regardless of their academic level. What is tangible and real to you may just be an idea, a concept, or a theory to them.

> Your concern for making things clear is shown through words and punctuation. The words *for example, namely, such as, in particular, specifically, and* the colon keep the attention of the reader at a high level because they promise easier understanding, less generalities, and more details.

Words alone, however, are often insufficient to bring full understanding. **Numbers** make adjectives real. **Visuals** — namely graphics, diagrams, tables, charts, and photos — help to clarify, analyse, explain, illustrate, and synthesise. Without visuals, a paper soon becomes unclear; without clear understanding, readers' attention soon wanes. Watch the frown disappear from the face of your reader when the words "shown in Fig. X" appear in your paper.

Question to Engage

Do you know what the most efficient attention-getter (and the best one to move ideas forward) is? It is, unfortunately, the most underused and underrated tool in the writer's toolbox. It is universal. It transcends languages. It guides the reader, triggers thought processes, and generates strong expectations. This attention-getter is . . . **the question**.

---

1. A question refocuses and prepares the mind.
2. A question challenges the mind. It cannot be ignored.
3. A question establishes the issue of a paragraph clearly.

What method provides enough contact force to polish these highly complex surfaces? Manual polishing with a belt machine would appear to be the obvious answer.

Take note of the clever way the expectations of the reader are set in the previous example. This "*obvious*" widespread technique may not be the best one, or the only one.

Professor Wolynes clearly loves questions too. He uses them to warm up the reader's mind to a new idea, away from the conventional one.

"Instead of unidirectional motion along a single pathway, can we have unguided motion through the myriad of shapes? Asking this question leads us in the right direction. We are forced to envision all the possible structures of the protein and how they are arranged and connected."[a]

One might think that all questions come with a question mark. This is not so. Adjectives, adverbs, and verb auxiliaries are often

---

[a] Reprinted excerpt with permission from Wolynes PG, "Landscapes, funnels, glasses, and folding: from metaphor to software", *Proc Am Philos Soc* **145**: 555–563, 2001.

questions in disguise. Here is Professor Wolynes again:

> *"Thinking in terms of energy landscapes, the Levinthal argument is **quite strange**."* [b]

> *"The energy landscape/funnel metaphor leads to a **very different** picture of the folding process than the pathway metaphor."* [c]

The reader is left wondering what makes the Levinthal argument strange, or how different the landscape metaphor is from the pathway metaphor. *"Quite strange"* and *"very different"* make such strong claims that they act as questions.

Recreate Suspense

The structure of a scientific article leaves little room for suspense. The gist of the contribution is revealed immediately in the title and in the abstract, well before the reader reaches the conclusions. Therefore, suspense has to be recreated. Questions excel at recreating suspense, but there are other ways. Sometimes, words announce an unexpected turn or show facts in a new intriguing light.

The following events will intrigue the reader.
1. A noteworthy contradiction, difference, exception, limitation: *however, but, contrary to, although, in contrast, on the other hand, while, whereas, whilst, only.*
2. An unexpected fact: *interestingly, curiously, surprisingly, the problem is that, should have (but did not), might have (but did not), unexpectedly, unforeseen, seemingly.*
3. A new alternative to go beyond the obvious: *rather than, instead, alternatively.*

---

[b] *Ibid.*
[c] *Ibid.*

*"**Although** COBRA (Cost Based operator Rate Adaptation) has shown itself to be beneficial for timetabling problems, Tuson & Ross [266, 271] found it provided only equal or worse solution quality over a wide range of other test problems, compared with carefully chosen fixed operator probabilities."* [d]

In the next example, the modal verb *"might have"* intrigues the reader.

*The Global Induction Rule method [3], a natural language processing method, **might have** worked on news video segmentation since news contents can be expressed in a form similar to that used for text documents: word, phrase, and sentence.*

...*might have, but did not!* "*Might have*" sets the expectation that the writer will explain why the method is not as applicable as originally thought.

In the final example of this chapter, observe how the author sustains the interest. In four consecutive sentences, he brings (1) an example, (2) two numbers, (3) a figure, (4) the attention-getters "however" and "important contradiction," and (5) a question suggesting one of the reasons for a difference in results.

For example, Strunfbach (6) reported a 27% increase in error rate when using the annealing method to improve the initial clusters obtained by the Clusdex method. Using the same methods and the same data, we observe a 52% decrease in error as seen in Fig. 3. In our case, however, cleaned and normalised data is used instead of cleaned data only. We therefore need to evaluate whether our findings represent an important

---

[d] Reprinted with permission from Sinclair M, PhD thesis, "Evolutionary algorithms for optical network design: a genetic-algorithm/heuristic hybrid approach", 2001.

contradiction. In particular, we need to ask: Are our data normalisation assumptions valid?

In this chapter, we examined many ways of enhancing and sustaining the initial attention of motivated readers, ways to help them go past the parts of a paper where attention is being drained at a fast pace. We used words, punctuation, syntax, text style, page layout, structure, examples, summaries, and visuals. Since readers' attention is not deserved, but earned through hard work, it is time to practise.

Read your paper to identify the parts the reader might find hard to understand (give it to read to somebody else if you are unsure). In these parts, modify your text accordingly to increase attention and facilitate understanding (examples, visual aids, questions, new subheadings, etc.).

Disclaimer: Attention-getters are only effective if used sparingly. The author of this book will not be responsible if excessive use of attention-getters in a paper distracts readers and makes them lose focus. Examples of such excessive use include writing *"importantly"* seven times in a long paragraph, turning a paper into a cartoon through overabundant visuals, using bold and italic text in so many places that the paper looks like a primary school paper, or starting three consecutive sentences with *"however"*.

3

Reduce Reading Time

Feeling time

In each scientific skills seminar I conduct, each participant is given a whole chapter of a book to read — 11 pages in total. It is to be read in about 20 minutes. Yet, after reading about six or seven pages, a few restless readers will flip pages forward to see how much reading is left. Do they need to see the light at the end of the tunnel? Do they tire because they require greater concentration to read a foreign language (i.e. English)? Or do they tire because they read methodically, progress slowly, and analyse each sentence?

Readers experience the passage of time differently for different reasons: familiarity with the topic, linguistic skills, reading motivation, reading habits, the writer's skills, or tiredness.

Visual Information Burgers

Scientific readers favour visuals — fast food for the brain, "information burgers" that reduce reading time and increase the informative value of an article. Visuals are more appealing than paragraphs of linear text because their message reaches the brain using very **high information bandwidth**. Whereas for text or speech, the brain processes one word at a time; for images, graphics, tables, and charts, the highly developed visual cortex processes them fast and globally, not one dot at a time. The eyes rapidly scan, detecting patterns as they go. Should you need convincing, conduct the following experiment.

Which of the following two representations gives you the largest amount of information in the shortest amount of time: the following text

In our experiment, for an upflow velocity of 0.10 m/h, the observed normalised tracer concentration of the effluent increased rapidly from 0 to 0.4 after 15 hours. The increase slowed after 38 hours when the concentration reached 0.95. It peaked at 1.0 at 90 hours, following which the concentration curve decreased steeply down to reach zero asymptotically at 180 hours. The calculated data and the observed data were closely related. However, when compared with the calculated data, the observed data seemed to lag when the concentration dropped.

or the figure in ☞1?

Separating Space

Space acts as a separator. It speeds up navigation to various parts of the paper. Because of space, *titles, headings, and subheadings* are quick to find and to read. Because the white background around them acts as an "eye trap", they receive greater attention. Short and simple in syntax like the heading above this paragraph, they are easily understood by the reader.

Indentation, by adding space in front of a paragraph, helps the reader rapidly locate start and end points (☞2).

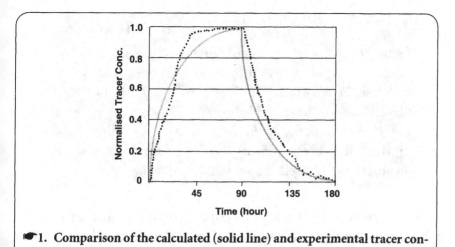

☞1. **Comparison of the calculated (solid line) and experimental tracer concentrations for an effluent velocity of 0.10 m/h.**

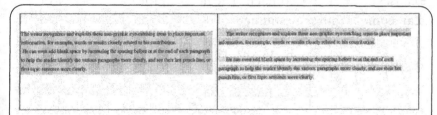

☞2. **White space between paragraphs.** Ignore the small print; it is the spacing that matters here. Compare the layout in the two boxes. Does the space between paragraphs help the reader?

The white space surrounding structural elements (headings, subheadings, visuals, formulas) also speeds up navigation inside the paper and allows the reader to rapidly focus on areas of interest, thereby saving time (☞3).

Trimmed and Discarded Text

Blaise Pascal

The 17th century French mathematician and philosopher Blaise Pascal makes an apology to a reader in
(Continued)

> *(Continued)*
>
> *Lettres provinciales.* He apologises because his letter is long, a lot longer than his previous ones. "This letter is longer", he writes, "only because I have not had the leisure of making it shorter." Conciseness seen as politeness! Boileau, a French writer from the same century, has harsh words: "Who knows not how to set limits for himself, never knew how to write."

As Pascal points out, a lengthy paper takes less time to write than a short one. Identifying the sources of excess length at a global level is the first step towards conciseness, but it is as difficult as determining the causes of a bulging stomach. The need for the diet is clear, but the fat can come from many sources.

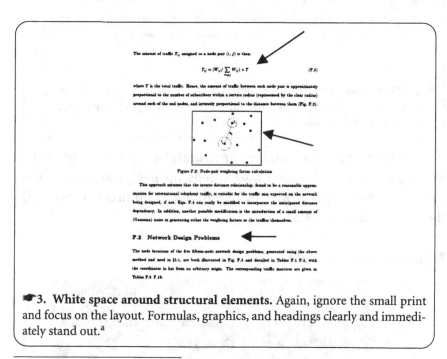

☛3. **White space around structural elements.** Again, ignore the small print and focus on the layout. Formulas, graphics, and headings clearly and immediately stand out.[a]

---

[a] Reprinted with permission from Sinclair M, PhD thesis, "Evolutionary algorithms for optical network design: a genetic-algorithm/heuristic hybrid approach", 2001.

1. Length is caused by the thousand words that should have been a diagram, a graph, or a table.
2. Length grows out of a structure still in the formative stage, where information is needlessly repeated in different sections.
3. Length is born out of the slowness of the mind, as it warms up and spreads a fog of platitude, particularly in the first paragraph following a heading or subheading.
4. Length is the fruit of unrealistic writer ambitions, aiming at cramming in a single paper the contribution of several papers.
5. Length is the fruit of hurriedness, since it takes time to revise a paper for conciseness.
6. Length happens when the reader is given details of the unnecessary kind, details unfiltered through the sieve of the contribution.

Is it possible to be too concise? Yes, if cutting time short makes your contribution unclear or difficult to assess. Here are four good reasons to justify lengthening a paper.

1. Lengthen to write a longer introduction that really sets the context and highlights the value of your contribution. Your contribution is like a diamond: to hold and display it, you need a jewel box, not a matchbox. Your introduction needs to motivate your readers.
2. Lengthen to repeat aspects of the contribution in every section of the paper (different levels of detail and different words). Each facet of a diamond contributes to its sparkle. Likewise, each part of a paper shows the same contribution at a different angle.
3. Lengthen to go beyond stating results in the abstract, and indicate what potential impact your contribution has on science. Would you give an uncut diamond and ask readers to polish it themselves?
4. Lengthen to provide the level of detail that enables research colleagues to independently assess your results and follow your logic.

The story at the head of this chapter demonstrates that time is felt differently by different people. People new to your field may feel it more than old-timers. For the newcomers, paradoxically, a longer introduction reduces the overall time required to read your paper because it sets the foundations to understand the rest of the paper. Experts can always skip or read the known parts in a cursory fashion.

In conclusion, you can control how the reader perceives time by (1) managing the length of the sections of your paper, and the number of titles and subtitles, according to anticipated reading difficulties; (2) speeding up reading through the use of visuals; (3) separating from plain text the elements used for navigation; and (4) removing needless words.

Read your paper. Are you repeating yourself? If you are, revise the structure. Do you feel that readers of the journal in which you publish your paper already know what the first paragraph of your introduction says? If you do, cut it out. Is the last paragraph of your introduction a table of contents for the rest of the paper? If it is, cut it out. Are you bored reading your own prose? If you are, it is time to replace it with a visual. Are you struggling to express the gist of your contribution in a few sentences only? Seek help, or "divide and conquer" and write several papers. Are all details essential to your contribution? Read the whole paper again and cut, trim, or discard.

4

Keep the Reader Motivated

Wrong title and unmet expectations

You are a young US researcher in the field of English speech recognition. You are still new in the field, and you are interested in general articles on how automated speech recognition is used in over-the-phone plane reservation systems. Searching through a large database of titles, you find the following article:

"Over-the-phone dialogue systems for travel information access."

You smile; all the keywords you typed are there. You order the paper through your library. A day later, it lands on your desk with a yellow post-it note attached to the first page that says, "A French girlfriend, maybe?"

Puzzled, you remove the post-it that covers the abstract and start reading. The abstract is at odds with the title. You had hoped for a general article, but you find that it is about French speech in dialogue systems. Your eyes

(*Continued*)

> *(Continued)*
>
> move to the name of the first author: Michelle Mabel. A French woman! Darn! No wonder the librarian is teasing you. Why else would you be reading an article so foreign to your research field? Should you start reading? Or should you worry about the rumour that is probably going around the lab about your torrid affair with a French woman? (To be continued.)

Dash or Fuel the Hopes of Your Readers: Your Choice

Motivating the reader starts with the title of your paper. It provides the initial reading impetus. The reader will scan hundreds of titles and select only a few. Imagine for a moment that the reader found *your* title interesting. You have what all authors dream of: the reader's attention. So it is now up to you. Are you going to dash your reader's hopes or, on the contrary, fuel them?

1. **Dash** — by a title that is not representative of the rest of the article.
2. **Fuel** — by a title that is representative of the rest of the article.

> You decide to read it anyway. After all, the paper is only five pages long. You should be able to get through it fast. You will just skip the parts that do not interest you.
>
> Half an hour later, you are only in the middle of page 2, reaching the end of the introduction, gasping for a graphic or a diagram to make things clearer. You glance anxiously at your watch. You have a meeting coming up with your team in 20 minutes. (To be continued.)

3. **Dash** — by making clear that reading the article will require more time than anticipated.

4. **Fuel** — by making clear that reading the article will require less time than anticipated.

> The introduction mentions French phonetics, and the differences in accents between the Chtimi and the Marseillais dialects. Your knowledge of France is limited to football players and perfumes. You have heard of Zinedine Zidane and Chanel #5, but that's about it. You flip pages to look at the reference section. No help there. You look up Wikipedia online. No help there either. If you ask the librarian, she will ask you about your French girlfriend, so you scrap that idea. By now, your motivation is at its lowest point, so you skip a few paragraphs and jump to the part about the dialogue modelling. (To be continued.)

5. **Dash** — by not giving the reader the baseline knowledge required to read the paper.
6. **Fuel** — by providing the reader with the baseline knowledge needed to read the paper.

> However, you did not know that your motivation could drop lower than its previous level. The key paragraph you found that seems to be precisely in your area of interest is totally obscure. You spend a good 5 minutes on it and then give up. (To be continued.)

7. **Dash** — by using prose so obscure or complex in syntax that the reader gets discouraged and becomes unsure whether he or she understands correctly.
8. **Fuel** — by using prose so clear that the reader is encouraged and sure that he or she understands correctly.

> You finally decide that the article is too specific. The semantic modelling will not apply to English at all. You won't be able to use it. And your meeting starts in a few minutes. A colleague who is also going to the meeting appears in your cubicle, looks quickly at one of the subheadings of the paper, and says, "I didn't know you were interested in French. Got a French girlfriend?"… You throw the paper in the trashcan. (To be continued.)

9. **Dash** — by making the reader doubt the quality, validity, or applicability of the contribution.
10. **Fuel** — By demonstrating to the reader the quality, validity, or applicability of the contribution.

> You rush to your meeting. As you enter the room, all your colleagues shout what sounds like "*Bonjour*". Before you answer, you look around and relax; your boss has not yet arrived. You say, "It is not what you think. The title was misleading." They all laugh. At that moment, your boss enters the meeting room and hands out a paper. "Here, read this", he says. "I have not read it, but it seems perfect for your research. It's by … um, a French name." The whole group collapses in laughter. (The end.)

11. **Dash** — by boring the reader with a style lacking dynamism, a sentence structure lacking variety, new information lacking emphasis, and text lacking illustrations.
12. **Fuel** — by captivating the reader with a dynamic style, a varied sentence structure, new information with emphasis, and text rich in stories and illustrations.

I deliberately chose a story to illustrate the various disappointments researchers experience while reading certain papers.

How much more dynamic is the language of stories compared to the stiff "classic" scientific writing style. Deviating from the norm is often frowned upon (like ending sentences with a preposition). Yet, one section of your paper is ideally suited to accommodate such deviations: the introduction. You have a story to tell: the story of why you embarked on your research, why you chose a particular method, etc. (see chapters 13 and 14 on the introduction). Since it is a story, use a story style to write it.

You can now see that your writing has much to do with sustaining the motivation of the reader through a combination of writing style, honest title, judicious detail and background, clear contribution, and good English.

Meet the Goals of Your Readers to Motivate Them

In our story, the reader (a newcomer to the field) is interested in general background. There are many kinds of readers, all coming to your paper with different motives and different levels of expertise. Satisfying and motivating them all is an impossible exercise if you do not really understand what readers hope to find in your paper. The following scenarios will help you understand their goals.

> ### The field intelligence gatherer
>
> Hi! I am a scientist working in the same area as you. I may not be doing the exact same research, but I am a regular reader of the journal you read and attend the conferences you attend. I was the guy sitting on the fifth row facing you when you presented your paper in Korea last year. I read most of the abstracts to keep up to date with what's happening.

First among the six reader profiles examined is the intelligence gatherer. Such scientists are interested in anything in an abridged form: your abstract or conclusions, sometimes the introduction. They probably will not read your whole paper unless they happen to be in the same field.

> **The competitor**
>
> Hi! You know me and I know you, although we have never met face to face. We reference each other in our papers. By the way, thanks for the citation. I am trying to find a niche where you are not playing, or maybe I'll fix some of your problems in my next paper. Hey, who knows, maybe you are onto something I could benefit from. I'd love to chat or work on a common paper one of these days. Interested?

Even if some of your background is missing, competitors are able to fill in the blanks without your help. They will read your whole paper rapidly, starting with the reference section to see if their name is in it and if their own reading is up to date. They may also use your list of references to complete their own list. Occasionally, a competitor may be asked to review your article prior to publication.

> **The seeker of a problem to solve**
>
> Hi! You don't know me. I am a senior researcher. I just completed a major project, and I am looking for something new to do. I am not quite familiar with your field, but it looks interesting, and it seems as though I could apply some of my skills and methods to your problems and get better results than you. I am reading your paper to find out.

Problem seekers may read the discussion, conclusions, and future work sections of your paper. Since their knowledge is not extensive, they will also read the introduction to bridge their knowledge gap.

> **The solution seeker**
>
> Help! I'm stuck. My results are average. I am pressured to find a better solution. I need to look at other ways of solving my problem. I started looking outside my own technology field to see if I could get fresh ideas and methods. I'm not too familiar with what you're doing, but as I was browsing my list of titles, I discovered that you are working in the same application domain as I am.

Solution seekers will read the method section, the theoretical section, and anything else that can help them. They could be surgeons looking for artery modelling software, or AIDS researchers who heard that small-world networks have interesting applications in their field. Their knowledge gap may be very large. They are looking for general articles or even specific articles, which they will read in part, expecting to find a clear and substantial introduction with many references to further their education.

> **The young researcher**
>
> Hello! I'm fresh out of university, and quite new to this field. Your paper looks like a review paper. That's exactly what I need right now. Nothing too complicated; just enough for me to understand the field, its problems, and the solutions advocated by researchers. That will do just fine!

Young researchers will read the introduction and (maybe) follow your trail of references. They do not expect to be able to make sense of everything the first time, but what little they can understand, they will be happy with. Their knowledge gap is great.

The serendipitous reader

Hi! Cute title you've got there. I had to read your paper. Such a title could only come from an interesting writer. I thought I would learn a few things, a paradigm shift maybe. I'm not sure that I will understand any of it, but it's worth a try. Last time I did that, I learned quite a lot. The paper had won the Best Paper Award in an IEEE competition. I studied the paper. Although I did not understand much, I got quite a few hints on how to improve my scientific writing skills!

My point is this: researchers will come to your paper with different motivations and needs. A common mistake is to imagine the reader as another you, the competitor in this story or someone who knows what you write about. As the author, you would be wise not to rush through the introduction and the list of references. You would also be wise to provide enough detail, so that other researchers can check and validate your work: "little validation, little value."

Ask people to read your paper. Ask for their opinion. Is it written for experts like yourself, or will researchers new to the field be able to benefit from your paper? Are they motivated to read the rest of your paper after reading your introduction?

5

Bridge the Knowledge Gap

Apple Computer

At Apple Computer (I worked there for 14 years), our director, Jean-Louis Gassée, used to compare people in the mainframe world with grand priests in white blouses serving unapproachable computer Gods. When the personal computer arrived, the Gods did not quite fall off their pedestals. They just moved from the computer room temple to the living room shrine. The computers had tamed their owners. Occasionally, through what looked like sorcery to humans, some PC owners managed to tame their computers. Application programs responded with docility to the secret incantations they typed. That knowledge known only to them had corrupted their virgin brains, and they no longer remembered what it was like *not to know*. Then came the Macintosh, to give hope again to the rest of the human race. History tells us that the Mac dispelled the Orwellian vision of 1984, but it did not quite manage to allay the fears of the uninitiated, although the iPod (another Apple product) did.

Crossing the knowledge chasm between you and your reader is not easy. The reader knows less than you. How much less?

1. It depends on you. If the new knowledge you are contributing is significant, the knowledge gap between you and your readers is large.
2. It depends on your readers. If their own knowledge of your field is small, they may not be familiar with the vocabulary or methods used. As a result, their initial knowledge gap is large, even if the additional knowledge you bring is modest.

You need to evaluate the gap to make sure your paper reaches the readers described in the previous chapter: the field intelligence gatherer, the competitor, the problem or solution seeker, and the young researcher. To reduce the uncertainty brought by the diversity of readers, one could assume that the reader is knowledgeable enough to follow your paper. Estimating the gap would then be simpler because readers' prior knowledge could be ignored. However, as scientists, we have to question assumptions. What do you, the writer, *know for sure* about your readers' initial knowledge?

You know that your readers find interesting one or several keywords in your title. You know that your readers are confident that they have *sufficient* knowledge to tackle and explore *parts* of your paper. You know that your readers read the journal your paper is published in or attend the conference where your paper is presented. Their work is related to the domain covered by the journal or conference. For example, readers who participate in the International Symposium on Industrial Crystallisation are in chemical engineering. They know the tools and techniques used in this domain. They know the meaning of centrifugation, phase separation, concentration, calorimetry, and polarised light microscope. They know the principles of science, how to conduct experiments, and how to read a concentration and temperature chart. They know English — the Queen's English, the President's English, but most probably some flavour of broken English.

Now that we have ascertained what you know for sure about your readers' initial knowledge, what then do you *not know for sure*? The answer is short and simple: **everything else**. Indeed, everything else cannot be assumed to be known. Even though it is tempting to believe that readers have the same level of knowledge as the one you had at the start of your project, nothing could be further from the truth. Readers are not younger versions of you.

Let us now consider your contribution. How much do they know about it? Nothing! Your contribution is unknown to them, **just as it was unknown to you before you started the research that led to your paper.**

Let us suppose that the title of your article is this:

"Phase transitions in lysozyme solutions characterized by differential scanning calorimetry."[a]

Some readers may be more familiar with characterisation techniques than they are with lysozymes. Therefore, they do not know which data, method, or experiment best applies to lysozyme, nor would they know what others before you have done in this domain or what specific problems remain unsolved. This is precisely what they will discover while reading your article.

'Ground Zero' Bridges

I hope you now see that, by and large, the gap between your knowledge and readers' knowledge is wide. Since it is impossible to

---

[a] Lu J, Chow PS, and Carpenter K, "Phase transitions in lysozyme solutions characterized by differential scanning calorimetry", *Prog Cryst Growth Characterization Mater* **46**(3):105–129, 2003.

guess how wide the gap is, you will have to decide how far back in time you will go to set a lower knowledge boundary, a reasonable "ground zero" on which you will build the knowledge of the reader.

To put it in an easy-to-remember formula:

Reader Knowledge Gap = *the new knowledge you acquired during your project* + *the new basic knowledge the reader requires to reach your ground zero.*

Even if many readers will read your article to educate themselves, it would be unreasonable to write your paper for college students or for scientists who are not the regular readers of the journal your paper targets. Ground zero will also be conditioned by the number of pages given to you by the journal. The more pages you are given, the more you can lower ground zero or increase the size of your contribution; for short papers, you will have to settle for a higher ground zero or a smaller contribution.

You could use the **reference section** of your paper as a "knowledge bridge". References are convenient shortcuts that, tell the reader, for example, *"If you want to know more about that, go and look at reference [1], [2], and [3]. I'm not going to explain about hidden Markov models. Indeed, I'm going to use the acronym HMM when I refer to them. This should be common knowledge to anyone working in the domain of speech recognition, his or her ground zero. Go and read reference [6], which represents the seminal work on the subject."*

Ground zero is often set by the latest books or review articles written by domain experts. If such articles are not available, then a look at the latest conference proceedings in your domain area should give you a couple of general articles that will accomplish the same function. Ground zero keeps moving up. Science is built on science, and scientists are expected to keep abreast of what is happening in their domain.

You may decide to lower the ground zero by providing extra background knowledge, instead of asking readers to get up to speed by themselves. Assuming the editor gives you enough pages, you can provide this knowledge in a **background knowledge section** that immediately follows the introduction. This section is a great place to summarise what readers would have learnt if they had had the time to read the articles that you put in your reference section (you can safely assume they will not read your [1], [2], and [3] before trying to read the rest of your paper). This background knowledge section is not part of your contribution, but it is necessary to understand it.

A word of warning, however. It is very tempting to go beyond a simple knowledge upgrade. If, in the background knowledge section, you go beyond providing the background and start presenting your contribution, make sure you identify clearly what is yours and what belongs to others. Beware of the passive voice: it makes the subject disappear and, as a result, the reader no longer knows who does what (authorship disappears). The journal editor must be able to assess your contribution clearly; therefore, you must clearly demark your contribution from that of others. Clearly state what is yours by using the pronouns *we* or *our*, sometimes accompanied by the present tense. What belongs to others is mentioned with or without the author's name, but with a reference and using the past tense (although the present tense may be used to strengthen other people's findings that you consider true in your paper).

Which of the following two sentences is the most convincing?

(1) *Tom* et al. *identified a catalyst that increases the yield at high temperatures.*
(2) *Tom* et al. *identified a catalyst that increased the yield at high temperatures.*

In (1), the present tense indicates that neither the catalyst nor the increased yield is in doubt. It would be incorrect to follow (1) with *"Slinger et al. showed that the increased yield is not due to the catalyst."* But, because the "*Slinger*" sentence uses the present tense, it can follow (2) to show that the author agrees with Slinger *et al.* and disagrees with Tom *et al.* One sees that the past tense does not have the convincing power of the present tense.

The Research Logbook: Keeping Track of the Knowledge Gaps

Your own research logbooks (in book form or online) are a gold mine of information, *if well kept.* Such logbooks come in handy when the time comes to write a paper. How do you keep them? Are some of their contents highlighted to mark a significant problem or discovery? Every discovery is a conceptual leap. Every leap separates you a little more from the reader. Can you afford not to mention that discovery in your paper? Will the reader understand your paper fully without it? Did you put "smiley" () emoticons in the margin to indicate your joy at having made a breakthrough, i.e. are your emotions captured? "*We were curious to find*" is a better phrase to motivate the reader than "*the problem was studied*". Will you be able to recapture the energy and fun of your research when, fingers on the keyboard, you type the first lines of your introduction? Will you remember these little bubbles of excitement that kept you motivated along the way; or is it the case that the project is finished, the fun is gone, the only thing that stands between you and your next research is that darn paper, and you are ready to tell the reader how frustrated you are having to write it?

In his excellent little book *A Ph.D. Is Not Enough*, Professor Feibelman gives sound advice:

> "*Virtually everyone finds that writing the introduction to a paper is the most difficult task. .... My solution to this problem*

*is to start thinking about the first paragraph of an article when
I begin a project rather than when I complete it.*"[b]

This guarantees that the introduction will not be stale.

To use a metaphor, imagine yourself standing next to your reader
at the foot of a hot air balloon. You are about to embark on your
research. You climb inside the gondola. As research progresses, the
landscape you see is different from the one seen by the reader on
the ground. Your discoveries raise your knowledge level as you drop
the sandbags of your ignorance. That ignorance now becomes your
reader's ignorance. Your balloon rises slowly (or rapidly, for one does
not usually dictate the pace of discoveries). By the time you are ready
to write, you have risen far above the reader, as in ☞1.

Your job as a writer is to bridge the gap created through months
of research. You have to throw a figurative ladder to readers, so that
they can come on board your hot air balloon gondola. How far down
should you throw the ladder? Down to ground zero. Oftentimes, the
ladder is too short. Ground zero is not properly identified — the
background is for tall experts only. Or, it may be that ground zero
is properly identified, but rungs in the ladder are missing. Readers
remain suspended in midair, frustrated, trying to get on board but
unable to climb further — you skipped an essential logic step that
prevents them from completely benefiting from your paper. Asking
a reader to be familiar with reference [X] before being able to understand the rest of your paper may be equivalent to asking the reader
to get off the ladder, go to the library, read *the whole article* [X], and
then climb up the ladder again. It would be more advisable, space
allowing, to briefly summarise in your paper whatever reference [X]
contains that is of interest to the reader.

---

[b] Feibelman PJ, *A Ph.D. Is Not Enough: A Guide to Survival in Science*, Basic Books, New York, 1993.

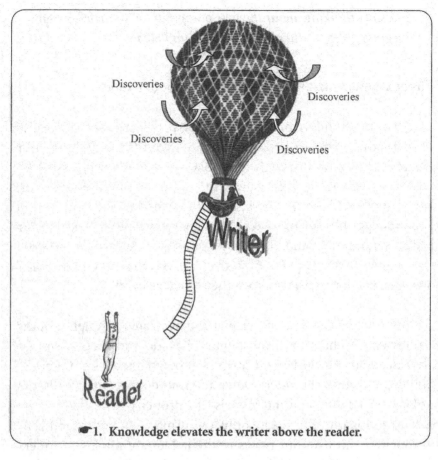

1. Knowledge elevates the writer above the reader.

The authors could have written this:

The dynamic behaviour was expressed in the Unified Modeling Language (UML; Booch et al. 1999). The notation used in Figure 3 is that of UML sequence diagrams. It is assumed that this notation is familiar to readers.

Instead, they wisely preferred to add a rapid overview of the notation in a footnote.

For those not familiar with the notation: objects line up the top of the diagram. An object's messaging and lifeline boundary is shown by a vertical dashed line starting below the object. Object activity is shown by the activation bar, a vertical rectangle drawn along the lifeline. Horizontal arrows issued from a sender object and pointed to a receiver object represent the messages sent.

Read your paper. Is your introduction too short? Is it motivating? Have you identified a ground zero that is reasonable to expect from your reader? Are you able, just from your research log, to identify the intermediate discoveries that removed the sandbags of your ignorance and elevated your knowledge above that of the reader?

6

Set the Reader's Expectations

To set an expectation is like creating a void that other sentences will fill. Not filling that void is tantamount to frustrating readers by not bringing closure. Controlling these expectations, limiting their range, or channelling them is not optional: it is key to the success of your paper. Therefore, manage the expectations you set.

Generally, expectations are set through the way a sentence is structured, its grammar, its syntax, its punctuation, and its words. Scientific expectations, on the other hand, are specific. Readers expect the author to bring evidence to justify scientific claims. They also expect the author to elaborate on what is new or unusual. Finally, they expect to have arguments and facts presented in a logical order: from hypothesis to observations, and from results to analysis and interpretation.

Expectations from Grammar, Syntax, Punctuation, and Words

In most languages, what comes at the end of a sentence is usually new to the reader (the sentence stress). Because the unknown is more interesting than what is known, we generally expect to find important information at the end of a sentence.

Main clause–subordinate clause

It is time to revive grammatical concepts acquired in secondary school. One recognises a main clause because it stands alone. A subordinate clause does not stand alone: to be understood, it needs the main clause. The following sentence has two clauses (each with subject and verb).

Learning needs to be semisupervised ←**main clause**
because variation within each class is large. ←**subordinate clause**

Learning needs to be semisupervised because of the large number of variations within each class. ←**no subordinate, one subject and verb only**

The next two sentences do not have subordinates. Their clauses are linked by the "*but*" conjunction of coordination. Which sentence seems more favourable to the evolutionary algorithms, (1) or (2)?

(1) *Evolutionary Algorithms are sufficiently complex to act as robust and adaptive search techniques; however, they are simplistic from a biologist's point of view.*

(2) *"Evolutionary Algorithms are simplistic from a biologist's point of view, but they are sufficiently complex to act as robust and adaptive search techniques."* [a]

Most of you find that sentence 2 is more favourable. The accumulation of favourable adjectives at the end of the sentence is convincing: *"sufficiently complex"*, *"robust"*, *"adaptive"*. Sentence 1 seems to put evolutionary algorithms in a mediocre light: the clause at the end of the sentence labels the algorithms with a pejorative adjective, *"simplistic"*. It would therefore seem that what ends a sentence influences your judgement more than what starts it.

The end of a sentence is not the only influential factor. In a complex sentence composed of a main clause and several subordinate clauses, the main clause is also main in the mind of the reader. Placed in a subordinate clause, information seems secondary, of lesser interest. Which sentence seems more favourable to the algorithms, (3) or (4)?

(3) *Although Evolutionary Algorithms are sufficiently complex to act as robust and adaptive search techniques, they are simplistic from a biologist's point of view.*

(4) *Evolutionary Algorithms are simplistic from a biologist's point of view, although they are sufficiently complex to act as robust and adaptive search techniques.*

In sentences 3 and 4, two phenomena combine their strength: the placement at the end of the sentence, and the position of the

---

[a] Reprinted with permission from Sinclair M, PhD thesis, "Evolutionary algorithms for optical network design: a genetic-algorithm/heuristic hybrid approach", 2001.

main clause in the sentence. It is highly probable that, in (3), you view the algorithms negatively; whereas in (4), you may hesitate to make a choice. When the main clause loses its position at the end of a sentence, it also loses its convincing power. In (4), the two influential factors are on opposite sides of the sentence (main clause at the head of the sentence and subordinate at the end of the sentence), so they neutralise one another. The reader will then decide according to personal preferences. The biologist will find them simplistic; and the computer programmer, sufficiently complex.

When asked how they viewed the evolutionary algorithms after reading the four sentences above, thirty-three scientists stated their preferences reflected in the last three columns of table (☞1). Note the role of the two main influential factors: main clause and end of sentence.

☞1. Influence on readers of information either contained in main clause or placed at end of sentence.

	Sent. start	Sent. end	Main	Sub.	Neg.	Equal	Pos.
Sentence 1	Pos.	Neg.	Pos./Neg.	...	15	14	4
Sentence 2	Neg.	Pos.	Neg./Pos.	...	3	4	26
Sentence 3	Pos.	Neg.	Neg.	Pos.	22	9	2
Sentence 4	Neg.	Pos.	Neg.	Pos.	12	19	2

Sent.: sentence; sub.: subordinate; pos.: positive; neg.: negative.

It is interesting to observe how language influences the way you interpret these sentences. If your original language is strongly influenced by Sanskrit or Pali, what others consider negative is positive to you, and vice versa. Your choices simply confirm the role of grammar in the way you analyse a sentence. Be aware that if you let your foreign grammar influence the way you write English, it may confuse the English-speaking readers as well as the Chinese and the Europeans.

> **Unscientific bias**
>
> All four sentences [(1), (2), (3), and (4)] present iden-
> tical facts: the algorithms are simplistic from one point
> of view, and they are sufficiently complex from another
> point of view. If you agree with one point of view
> (for example, if you find them simplistic), then all sen-
> tences 1 through 4 should carry the same message and
> be equally perceived, whatever the order of the words
> in the sentence. Yet, this is not the case, is it? You find
> some sentences favourable and others unfavourable. As
> you can see, the placement of words in a sentence is not
> neutral. It has a great influence on the way you perceive
> the facts presented.

Consider starting sentences with a subordinate clause in order to end them on a convincing main clause. The following sentence starts not only follow this principle, but they also make excellent attention-getters, they really shine at setting expectations, and they are fast to read.

If...	*Although...*
Since...	*Because...*
Given that...	*Instead of...*
When...	*While...*

Because the reader is familiar with a template-like grammatical structure (*if ... then, because ... then*), reading is faster.

Expectations from Science

Readers of scientific papers have different expectations than readers of novels. When the novelist writes "the ferocious dog", the reader is quite happy to imagine a ferocious dog and does not expect the writer to prove that the dog was ferocious. It may not be the same type of ferocious dog as the one the writer had in mind, but who cares — the more ferocious, the better! Unlike the scientist, the novelist does not have to convince the reader of the dog's ferocity by measuring the surface of the barred dental surface, the number of millilitres of saliva secreted per minute, or the dog's pupil dilation.

Setting expectations with adjectives and adverbs

Adjectives or adverbs are highly subjective. What is robust to you may be fragile to me. What is very fast to you may be moderately fast to me. In science, adjectives and adverbs are claims. The next two examples make adverbial/adjectival claims.

Traditionally, airplane engine maintenance has been labour-intensive.

This sentence makes two claims: "*labour-intensive*", and "*traditionally*". The reader expects that the author has found a way to make this process less labour-intensive. Why? The adverb "*traditionally*" indicates that so far (present perfect *has been*), this is the case, but things may be changing. In just one sentence, the author has indirectly stated the main benefit of his contribution (less labour-intensive) and prepared the reader to expect a contribution on robot-maintained engines or on maintenance-free engines. What expectation does this next sentence raise?

Up to this point, we have only considered basic filtration techniques.

The reader expects that more sophisticated filtering techniques will be presented. This expectation is raised through the adjective "*basic*". The locution "*up to this point*" closes one door and opens another. It is a great transition device.

Setting expectations with declarative statements

Do you remember the example with Tom Smith in chapter 1? The sentence is written in a passive voice: the subject (the "*assumption*") is not the doer of the verb ("*supported*"). Notice the preposition "*by*" that often accompanies the passive voice; it precedes the real doer — the "*data*", placed after the verb.

> *Tom Smith's assumption [4] that no top layer material could come from the byproducts of the pinhole corrosion which had migrated is not supported by our data.*

Because 2 subclauses and 17 words separate the subject from its verb, the sentence is unreadable. Here is the sentence again, but this time with verb closer to subject and with added dynamism through the use of the active voice. This sentence is written in the present tense, a tense usually used for claims. The auxiliary "*do*" shows certainty.

> (1)   *Our data reveal that, contrary to Tom Smith's assumption [4], the pinhole corrosion byproducts do migrate to form part of the top layer material.*

What is the expected topic of the next sentence? The data or Tom Smith? The majority of you will answer, "The data." You expect proof of the claim regarding migration, or proof that the material found on the top layer comes from the pinhole. An expectation is set, but Tom Smith's assumption did not set it. Tom Smith is now between commas, a side remark. The findings of the author are stated, not those of Tom Smith. If the author were to describe Tom Smith's assumption in detail after (1), then the reader, expectation unmet, would be frustrated. To create an expectation on the theme of Tom Smith, paragraph (2) would be more appropriate.

(2) *Our data reveal that the pinhole corrosion byproducts migrate to become part of the top layer material. These findings contradict Tom Smith's assumption [4].*

Tom Smith's assumption is no longer a side remark. It is the main point and it comes in a small package: a short punchy sentence. The reader is now curious. What did Tom Smith assume? Why is there a contradiction?

Setting expectations using the steps of the scientific process

In the following report, the author laid out the sentences in this paragraph according to a very specific order expected by the reader: that of the scientific process. Hypothesis precedes experiment, experiment precedes result, and result precedes discussion and conclusions. The four sentences are laid out in this specific order. The reader expects it.

*[**Hypothesis**] Since the dengue genome forms a circle prior to replication, alike the rotavirus and the polio virus, and since one end of the circling loop is at the 3′ end of the genome where replication takes place, <u>we wondered</u> if the loop had an active role to play in the replication. [**Experiment**] <u>After comparing</u> the RNA synthesis capability of various whole and truncated dengue genomes using radio-labelled replication arrays, [**Results**] <u>we found that</u> another region had an even larger role to play in the replication: the 5′ end of the genome. Although far away from the 3′ end, it seems to loop back into it. [**Discussion**] <u>Thus</u>, it may be that the promoter site for RNA synthesis resides in this unusual location. Looping would then be a means of bringing the promoter to where it can catalyse rapid duplication.*

Read your abstract and your introduction. Highlight all adjectives in fluorescent yellow, and adverbs in fluorescent red. If your paper glows in the dark, then you have got work to do. Examine each adjective and adverb. Are the claims justified? Would removing an adjective make you more authoritative? Could each adjective be replaced by a fact?

The second exercise is fun to do. Take ten consecutive sentences from your introduction. Look at each sentence one by one and try to guess what the following sentence contains without looking at it. If your guess is right, add two points to your total score. If you really cannot guess what comes next. Take one point away from your total score. If you guess wrongly, take away two points from your total score. If you score twenty, you are a brilliant writer who knows how to set the expectations of the reader. If you score less than twelve, consider rewriting your sentences to better control the expectations of your reader.

The first sentence of a paragraph often raises expectations that are answered in the rest of the paragraph. Check if this is so in your paper.

The last sentence of a paragraph often raises expectations that are answered in the paragraph following it. Check if this is so in your paper.

7

Set Progression Tracks for Fluid Reading

When readers cruise down your paper in fifth gear, it is because you have created a highway for their thoughts to travel on at great speed, a highway that stops their mind from wandering where it should not go. Sometimes, while reading some papers, I feel as if I am driving in the fog at a crawling speed across a muddy field trying to follow somebody else's tracks. In science, unlike literature, you guide your readers along a clearly lit, well-signposted highway. How to create such a highway is described in one word: progression.

Progression is the process of transforming what is new into what is known. It builds a coherent context that allows readers to travel light and read on with minimum cognitive baggage. When readers start a sentence, a paragraph, or a section in your paper, they relate what they read to what they know. This progressive anchoring of new knowledge onto old knowledge is an essential learning mechanism.

This chapter reviews two classes of progression schemes: topic-based progression; and non–topic-based progression such as progression by explanation, by logical step, or by transition word. Often times, these two schemes are at work together within the same sentence. In this chapter, they are considered separately to enable you to identify them more easily. Progression and expectation can work independently in the same sentence, but usually the expectations created in one sentence determine the type of progression in the next sentence. Topic-based progression relies on an understanding on what constitutes a topic.

Topic and stress

In France, schoolchildren discover all about topic and stress in their grammar book at the age of 14. By the time they go to university, they have forgotten almost everything. Topic and stress have a well-defined meaning in grammar. In a simple sentence, the topic corresponds to the known information located at the head of a sentence (often the subject of the main verb), and the stress is the new information located after the topic (the active verb and the rest of the sentence).

"But trapping" ←**Topic**
"leads to very slow dynamics at low temperature." [a] ←**Stress**

A complex sentence — containing more than one verb — may have more than one topic or stress. The topic is usually (but not always) a subject. To identify it, evaluate whether what it describes is known and is at the beginning of a sentence, or follows a colon or semicolon.

Learning, however, ←**Topic**
still needs to be semisupervised because variation within each class is large. ←**Stress**

For content-based medical image retrieval systems, ←**Topic**
the number of classes is limited. ←**Stress**

A sentence may have no topic at all; everything in it is new. Such a sentence is often found at the head of a paragraph, or where the author wants to surprise the reader by bringing (for effect and knowingly) the new information to the beginning of a sentence.

---

[a] Reprinted excerpt with permission from Wolynes PG, "Landscapes, funnels, glasses, and folding: from metaphor to software", *Proc Am Philos Soc* 145:555–563, 2001.

Three Topic-Based Progression Schemes to Make Reading Fluid

Progression around a constant topic

The scheme is straightforward. The subject of the sentence is repeated in successive sentences, either directly or through the use of a pronoun, a synonym, or a more generic or specific name. The reader is already familiar with the topic and reading is fluid. In this example, "*trapping*" is the constant topic.

> "**Trapping** is unimportant at high temperatures where there is plenty of energy to escape. But **trapping** leads to very slow dynamics at low temperature. In the case of liquids, this **trapping** causes the glass transition — a dramatic slowing of motion on cooling."[b]

In the first sentence, the author skilfully sets the expectation of readers to what comes next: an explanation of why trapping is important at low temperatures. This expectation is reinforced by the use of "*but*" for contrast. The second sentence also creates the expectation fulfilled in the third sentence that the author will give an example of "*very slow dynamics*", in this case "*the glass transition*". Note the use of an apposition, "*a dramatic slowing of motion on cooling*", which arrives just-in-time to explain the unusual keyword. In these three sentences, expectation and constant topic progression, although created separately, are acting jointly to make the text more interesting.

The first two sentences of the above text could have been written as follows:

> **Trapping** is important at low temperature because it leads to very slow dynamics, as there is not much energy for the molecules to escape.

---

[b] *Ibid.*

The total number of words for both texts is identical: 24 words. But the second text does not create the same compelling wish to know more about *"very slow dynamics"*. The expectation created by the second text is not as strong: the text is long and its two cascaded reasons (*"because"*, *"as"*) make it more complex. It is also less contrasted because low and high temperatures are no longer compared.

Progression through partial aspects or subclasses of the main topic

In this progression, the main topic is usually announced in the first sentence, and subsequent sentences examine aspects of the topic. In the following example, the first sentence is about visuals. The next two sentences review two aspects of visuals: their placement and their convincing power.

> <u>Visuals</u> *are star witnesses standing in the witness box to convince a jury of readers of the worth of your contribution.* **Their placement** *in your paper is as critical as the timing lawyers choose to bring in their key witness. More importantly,* **their convincing power** *is far beyond that of text exhibits.*

Chain progression

In a chain progression, topic and stress are daisy-chained. The stress at the end of a sentence becomes the topic at the beginning of the next sentence. This frequently used progression scheme is easy for readers to follow. It is illustrated in the next paragraph.

> **"The protein** *when it is first made exists in an extraordinarily large variety of shapes, resembling those accessible to a* **flexible strand** *of spaghetti. The Brownian motion of* **the protein strand** *will carry it willy-nilly between various shapes, somehow finally getting it to settle down into a much less diverse family of shapes, which we call the '* **native structure**' *of the protein. The*

*average **native structures** of many proteins have been inferred experimentally using X-ray crystallography or NMR."*[c]

The elements in a daisy chain do not need to be repeated word for word from one sentence to the next. Often, the verb in a previous sentence becomes a noun at the head of the next sentence.

*Applying Kalman filters **reduced** the noise in the data sent by the low-cost ultrasonic motion sensors. **The reduction** was sufficient to bring down the detection error rate below 15%.*

Sometimes, part of the previous sentence (as underlined in the following paragraph) is rephrased briefly at the beginning of the next sentence. When this happens, the sentence often starts with "*This*" or "*This result*".

*"The above observations can be generalized to a rather important conclusion. If large mole differences between species exist in a data set (and **this** is often the normal case for catalytic reactions), then the reactions involving both major and minor species should be rewritten to include only the latter. **This** should solve the problem of abnormal gradients in the extent of reactions for most cases."*[d]

Observe how the authors capture the readers' attention with words of importance ("generalized to a **rather important** conclusion") or words attractive to scientists "(**solve the problem** of abnormal gradients", "**generalized** to a rather important **conclusion**"). On the other hand, here is a wonderful example of an ambiguous pronoun. Consider the final "*This*". What does "*This*" represent? If you

---

[c] *Ibid.*

[d] Reprinted from Widjaja E, Li C and Garland M, "Algebraic system identification for a homogeneous catalyzed reaction: application to the rhodium-catalyzed hydroformylation of alkenes using *in situ* FTIR spectroscopy", *J Catal* **223**:278–289, 2004 (with permission from Elsevier).

answer "*the reactions*", then read the sentence again because it is not the correct answer. The final "*This*" refers to the rewriting.

Finally, let us establish a link between thematic progression and expectations. The constant topic progression answers the need to know more about it (expectation of elaboration — breadth). The derived topic progression answers the need to go deeper into the topic (expectation of elaboration — depth). The linear progression answers the need to see how things are related (expectations of relatedness and outcome). Non-thematic progressions have even more direct ties with expectations, as we shall see.

Non–Topic-Based Progression Schemes

Progression through explanation

The second sentence of the following paragraph introduces a progression of a new type: the **explanation**. It usually follows a question or a statement that acts as a question. The sentences in the example are numbered to facilitate its analysis.

(1) *Why are such discontinuities in progression so common in your first draft?* (2) *Because when you write a new sentence, thoughts relevant to previous sentences bubble up to the surface of your consciousness to disrupt smooth topic progression.* (3) *It is as if, while you are writing a sentence, your brain simultaneously launches a neuronal search party whose task is to retrieve related information.* (4) *As each search party asynchronously returns with its findings, it interrupts your writing with thoughts relevant to past sentences.*

Sentence 1 is explained by sentence 2. Sentence 2 is explained by sentences 3 and 4.

Progression by explanation is often announced by transition words: *for example, thus, indeed.* The word *similarly* helps progression by announcing an explanation based on an example, an analogy, or a metaphor.

What! Metaphors?

These last two chapters seem to discourage calling on the reader's imagination. One must define ferocity. One must keep the reader on track, channel thoughts, etc. Does this mean that you, the scientist, should keep imagination at bay on the grounds that science is objective? Many examples in this book come from an article written by Professor Wolynes entitled "Landscapes, funnels, glasses, and folding: from metaphor to software". Rather than trying to convince you, I will quote here the first few sentences of his article: "Of all intellectuals, scientists are the most distrustful of metaphors and images. This, of course, is our tacit acknowledgment of the power of these mental constructs, which shape the questions we ask and the methods we use to answer these questions."[e]

Time-based progression

Time-based progression is the most common type of sequential progression.

(3) *It is as if, while you are writing a sentence, your brain simultaneously **launches** a neuronal search party whose task is to retrieve related information. (4) As each search party asynchronously **returns** with its findings, it interrupts your writing*

---

[e] Reprinted excerpt with permission from Wolynes PG, "Landscapes, funnels, glasses, and folding: from metaphor to software", *Proc Am Philos Soc* **145**:555–563, 2001.

*with thoughts relevant to past sentences. (5) This richly chaotic activity of your brain is part and parcel of the creative writing process. (6) This natural process is best left alone when you write your first draft. (7) Leave to **later** revisions the reordering of your sentences **after** you have decided whether to weed or to keep these disruptive thoughts.*

Between sentences 3 and 4, a new progression is at work: the **time-based progression**. First, the "*neuronal search party*" goes to explore the brain (3), and then it returns with information (4). Between sentences 6 and 7, the progression is also temporal: the writer goes from the "*first draft*" to the next version or "*later revisions*".

Although words such as *first, to start with, then, after, up to now, so far, traditionally, finally,* and *to finish* mark the start, the middle, or the end of a time step, time is often implicit. The scientific reader understands that the writer is following the logic of time when narrating the various steps of an experiment. Most often, the passage of time is established by changing the tense of a verb, from the past to the present or from the present to the future.

In a previous example, a time-based progression co-occurs alongside a chain progression. The tense changes from the present to the future.

*"The protein when it is **first made** exists in an extraordinarily large variety of shapes, resembling those accessible to a flexible strand of spaghetti. The Brownian motion of the protein strand will carry it willy-nilly between various shapes, somehow **finally** getting it to settle down into a much less diverse family of shapes, which we call the 'native structure' of the protein."*[f]

---

[f] *Ibid.*

Logical sequential progression

Enumeration is a type of sequential progression: from the first to the last. In one of the previous examples (*"Why are such discontinuities in progression so common in your first draft?"*), sentences are numbered and examined one by one.

Progression can be numerical, but it can also follow an order defined by the author (the elements of a list, for example). In the next sentence, the author announces two factors that contribute to the propagation of dengue fever before covering each one in turn.

Two factors contribute to the rapid spread of dengue fever: air transportation and densely populated areas.

Sometimes, the list is not explicit. The author will cover each noun that makes up a compound noun. In the following example, the compound noun is the *"dengue virus"* (*"dengue"* is the disease, and *"virus"* is the microorganism).

*The **dengue virus** from a human carrier is transmitted to the female Aedes mosquito that feeds on an infected blood meal. The **virus** multiplies inside the mosquito over 3 to 5 days. It is transmitted back into a human through the saliva injected by the mosquito when it bites. **Dengue** usually spreads because of human travel (particularly air travel), ineffective mosquito control methods, and poor sanitation in areas with water shortages.*

Progression through transition words

Progression is sometimes announced by special words called transition words, such as *in addition, moreover, furthermore, and, also, besides, first, then,* or *now.* These words are a topic of controversy among writers. Such transition words, some say, are just a convenient way to ignore progression: they artificially establish a transition where none exists. Actually, this is often, but not always, the case. I recommend that when you see these transition words, you try to replace them with an implicit progression such as a sequential step or a topic progression. If you cannot replace them, it may well be that an explicit progression using these transition words is necessary. The following long paragraph has 97 words. The revised paragraph has no transition word (in bold in the original) and has only 62 words.

> *[**Original**] "Formation of prognostic knowledge is concerned with extracting knowledge from historical data in a maintenance and diagnostic system. Different prognostic approaches can be used based on the characteristics of the equipment and the nature of the failure problems. For example, in cases where complete knowledge of the equipment is not available, it will be difficult to produce a comprehensive model for model-based prognosis. However, it is possible to use experience-based or AI-based approach [6] to extract the prognostic knowledge. **Furthermore** it is possible to gradually build the comprehensive prognostic system by combining the results of multiple approaches."*[g]

> *[**Revised**] When complete historical data from a maintenance and diagnostic system are available for all failure types of a particular piece of equipment, comprehensive prognostic knowledge can be formed. When little historical data are available, gradual knowledge extraction methods — experience-based or AI-based [6] — are necessary. Combined with the model-based method, these methods also enable the gradual building of a comprehensive prognostic system.*

In the revised version, the first two sentences cover two aspects of the same topic: complete historical data and incomplete historical

---

[g] Zhang DH, Zhang JB, Luo M, Zhao YZ, and Wong MM, "Proactive health management for automated equipment: from diagnostics to prognostics", *Proceedings of Eighth International Conference on Control, Automation, Robotics and Vision (ICARCV2004)*, Kunming, China, pp. 479–484, 2004. © 2004 IEEE.

data. The third sentence is in chain progression with the second sentence. The transition word has vanished. The contrast brought by the word "*however*" in the original version is not lost; instead, it is now established by the opposition between "*little historical data*" and "*complete historical data*". Parity of performance between the two systems is announced by the word "*also*" in the last sentence. The restructuring increases clarity and conciseness.

Faulty Progression and Pause in Progression

Sometimes, the author pauses to let the reader catch up. A summary, a restatement, a comment, or an example consolidates the reader's understanding. Words that announce a pause include *to summarise, briefly put,* and *for example,* to name a few.

Sometimes, the progression does not pause; it breaks. It becomes jerky, stops for a sentence or two, and then resumes its pace. In such situations, the reader rapidly loses his or her sense of direction. Somewhere, somehow, one or two links in the progression chain are broken, but where? The broken links are easily identified. Underline the topic of each sentence in a paragraph. Circle the topics that are not part of a topic-based progression (i.e. not connected to the topic or stress of the previous sentence). See if these topics are in a non–topic-based progression (explanation, time, or logical sequence). If they are not, then congratulations, you have just located a broken link.

(1) *After conducting microbiological studies on the cockroaches collected in our university dormitories, we found that their guts carried staphylococcus, members of the coliform bacilli, and other dangerous microorganisms when outside of the intestinal tract.* (2) *Since they regurgitate food, their vomitus contaminates their body.* (3) *Therefore, the same microbes, plus moulds and yeasts, are found on the surface of their hairy legs, antennae,*

and wings. (4) To find such microorganisms in their guts is not surprising, as they are also present in the human and animal faeces on which cockroaches feed.

The {topic | stress} pairs are as follows: (1) {cockroaches | microorganisms in guts}, (2) {cockroaches | regurgitation and vomitus contamination of body}, (3) {microorganisms | parts of body}, and (4) {microorganisms in guts | faeces}.

Sentences 2 and 3 cannot be separated because they are linked by progression based on a cause-to-effect explanation. Sentences 1 and 4 should be juxtaposed in chain progression, strengthened by a logical progression (effect-to-cause). Here is the improved paragraph.

After conducting microbiological studies on the cockroaches collected in our university dormitories, we found that their guts carried staphylococcus, members of the coliform bacilli, and other dangerous microorganisms. To find such microorganisms in their guts is not surprising, because they are also present in the human and animal faeces on which cockroaches feed. Since cockroaches regurgitate food, their microorganism-laden vomitus contaminates their body. Therefore, the same microbes, plus moulds and yeasts, are found on the surface of their hairy legs, antennae, and wings.

Sentence 3 is more logically connected to sentence 2 with the addition of "*microorganism laden*".

A broken link is often the consequence of an inversion between topic and stress. Why is this inversion a problem? Read the following sentence.

The cropping process should preserve all critical points. Images of the same size should also be produced by the cropping.

This paragraph does not seem well balanced, do you agree? This is because, in the second sentence, the already known information ("*the cropping*") is at the end, a place traditionally reserved for new information. Here are three ways to correct the problem:

1. Change the voice in a sentence from active to passive or vice versa, thus straightening the inverted topic and stress by bringing the known information to the head of the sentence.

    The cropping process should preserve all critical points. It should also produce images of the same size.

2. Invert the order of the sentences to re-establish progression.

    Images of the same size should be produced by the cropping. The cropping should also preserve all critical points.

3. Combine the two sentences into one.

    The cropping process should preserve all critical points and produce images of the same size.

Now that you are equipped to solve inverted topic–stress situations, discover another method. It requires a table in which you write down the topic and stress for each sentence, as well as the type of progression (☞1). The table is used in an example that illustrates and remedies an inversion problem (☞2).

The next paragraph is about a tropical and subtropical disease called dengue fever. <u>Knowing that it will be followed by another section describing how dengue can be prevented and controlled</u>, compare the original version with the final version, and identify how the text was improved.

☛1. **A method to detect progression problems in paragraph text.** This method identifies the topic, stress, and progression type of sentences. Progression schemes are as follows: (1) constant topic — the topic of successive sentences remains the same; (2) chain progression — the theme at the end of a sentence becomes the topic at the start of the next sentence; (3) subtopic — the main topic/stress appears in the head sentence, and other sentences dwell on aspects of it; (4) sequential step — from one sentence to the next, something has progressed to the next logical or time step; and (5) explanation — a sentence is explained or illustrated in subsequent sentences.

Sentence in paragraph	Topic(s) [known info at head of sentence, or the subject(s) of a sentence]	Stress [new info at end of sentence, or verb(s) and their object(s)]	Progression type (constant topic, chain progression, subtopic, sequential step, explanation)
1			
2			
3			
...			

☞**2. The new-before-old progression error.** The original has problems. Information 2(b) is already known in sentence 1(a), and stress and topic are inverted because of the passive voice. Three revised versions are proposed. In **(A)**, a constant topic progression around the nucleosome is re-established by using the active voice in sentence 2, and by inverting main and subordinate clauses in sentence 3. In **(B)**, sentence 1 is reorganised to consolidate the information on chromatin in one sentence (instead of two) and to establish a chain progression around the histones. Both (A) and (B) create the same expectation: how does a missing histone perturb the function of the nucleosome, and what happens because of it? In **(C)**, a constant topic progression, as in (A), sets a different expectation: the reader now expects to discover either other functions of the nucleosome or how it remodels the chromatin.

Original sentences: The nucleosome, a structural unit within the chromatin, has a length of DNA coiled around eight histones. The chromatin structure is remodelled by the nucleosome. But, if any of its histones are missing, the nucleosome may malfunction.

Sentence in paragraph	Topic(s) [known info at head of sentence, or the subject(s) of a sentence]	Stress [new info at end of sentence, or verb(s) and their object(s)]	Progression type (constant topic, chain progression, subtopic, sequential step, explanation)
1	(a) *The nucleosome*	(b) *a length of DNA coiled around eight histones*	
2	(a) *The chromatin structure*	(b) *remodelled by the nucleosome*	?
3	(a) *But if any of its histones*	(b) *are missing the nucleosome may malfunction*	?

Modified sentences:
(A) The nucleosome, a structural unit within the chromatin, is composed of a length of DNA coiled around eight histones. The nucleosome complex is important in remodelling the chromatin structure. The function of this complex may be disrupted when any of its histones are absent.
(B) The nucleosome is a structural unit within the chromatin, which it helps remodel. It is composed of a length of DNA coiled around eight histones. If any histone is missing, the nucleosome's function may be disrupted.
(C) The nucleosome, a structural unit within the chromatin, has a length of DNA coiled around eight histones. It may malfunction if any of its histones are missing. The nucleosome remodels the chromatin structure.

Original text

The transmission of the dengue virus to a human occurs through the bite of an infected female Aedes mosquito. In addition, the disease spreads rapidly in densely populated areas because of the lack of effective mosquito control methods, the increase in air travel, and poor sanitation in areas with a shortage of water. The mosquito becomes infected when it feeds on a blood meal from a human carrier of the virus. The virus multiplies inside the infected mosquito over 3 to 5 days, and resides within its salivary gland.

Follow these steps to analyse the original text.

1. Identify the author's intention, i.e. the main point of the paragraph (in our case, the title is revealing).
2. Isolate the key point(s) put forth by the author. Identify closely related sentences by the words they share. Identify potential sentence topics — usually the words repeated from one sentence to the next.
3. Identify a first topic on which to base the progression scheme, and start ordering the points in the paragraph (take into account the expectation you have to set for the next paragraph or section).
4. Restructure the text to establish progression and the desired expectations.

Solution

1. Author's intention
   The author first presents the cycle of propagation of the virus, because it is directly linked to how the spread of the disease can be prevented and controlled. It is critical to respect the correct sequence: transmission cycle, then propagation.

2. Key points
    The key points here are the human–mosquito–human transmission cycle, and the causes for the spread of the disease.
3. Topic and progression scheme
    It is essential to finish on the theme of propagation, because the author will continue with another paragraph on what the community can do to prevent the rapid spread of dengue.

Final text (version one)

The dengue virus from a human carrier is transmitted to the female Aedes mosquito that feeds on an infected blood meal. The virus multiplies inside the mosquito's salivary gland over 3 to 5 days. It is transmitted back into a human through the saliva injected by the mosquito when it bites. The virus spreads rapidly in areas where large numbers of humans and mosquitoes cohabitate. This spread is aggravated by human travel (particularly air travel), ineffective mosquito control methods, and poor sanitation in areas with water shortages.

In this version, the topic of each sentence is the same: the virus, or the disease caused by the virus. Progression is therefore built around a constant topic. The progression is also a time-based progression (the transmission cycle) and a logical progression (amplification: from limited to extended, from specific to general).

Final text (version two)

The female Aedes mosquito feeds on the infected blood of a human carrier of the dengue virus. Inside the mosquito's salivary gland, the virus multiplies over a period of 3 to 5 days. When the mosquito bites, its saliva carries the virus back into another human. In communities where large numbers of humans and mosquitoes cohabitate, the dengue virus spreads rapidly. This spread is aggravated by human travel (particularly air travel), ineffective mosquito control methods, and poor sanitation in areas with water shortages.

In the second version, the mosquito is the constant topic in the first three sentences. Using known information (mosquito, human, virus), the fourth sentence transitions gently to a new theme: propagation. The last sentence is in chain progression with the preceding one.

4. Text restructuring

Both versions have more or less the same length as the original. They give a better description of the human–mosquito–human virus transmission cycle. Note the transition sentence in the middle of the paragraph. It allows smooth logical progression between transmission cycle and propagation, and it prepares the chain progression with the last sentence.

If you have enough stamina and energy left after this exercise, you could try your hand on the following paragraph. It should be familiar to you, as it was corrected earlier in this chapter.

*[**Original**] "Formation of prognostic knowledge is concerned with extracting knowledge from historical data in a maintenance and diagnostic system. Different prognostic approaches can be used based on the characteristics of the equipment and the nature of the failure problems. For example, in cases where complete knowledge of the equipment is not available, it will be difficult to produce a comprehensive model for model-based prognosis. However, it is possible to use experience-based or AI-based approach [6] to extract the prognostic knowledge. Furthermore it is possible to gradually build the*

comprehensive prognostic system by combining the results of multiple approaches."[h]

1. Author's intention
   The author wants to present a method that allows the building of a comprehensive prognostic system, even when the information about the system to model is missing. This intention is clearly indicated by the adjectives *"comprehensive", "complete", "gradually",* and *"available".*

2. Key points
   Expert systems, jointly with other knowledge extraction methods, are able to progressively improve the prognostic models.

3. Topic and progression scheme
   The order is clear because the situation changes according to the availability (total or partial) of data to build the prognostic model.

4. Text restructuring

   > *[**Final**] When complete historical data from a maintenance and diagnostic system are available for all failure types of a particular piece of equipment, comprehensive prognostic knowledge can be formed. When little historical data are available, gradual knowledge extraction methods — experience-based or AI-based [6] — are necessary. Combined with the model-based method, these methods also enable the gradual building of a comprehensive prognostic systems.*

   One final word of caution: do not attempt to "fix" progression problems in a paragraph without taking into account the topic of the next paragraph. Progression applies between paragraphs just as much as it applies between the sentences of a paragraph. Progression

---

[h] *Ibid.*

problems are not always fixed by moving sentences around without much modification. In many cases, an unclear text needs complete restructuring prior to applying progression schemes. To restructure, it is indispensable to understand the author's intention and to identify the key point of the argument made.

Take ten consecutive sentences from the discussion or the introduction of your paper. Identify the topic and the stress of each sentence. Can you identify a progression scheme? Are some sentences totally disconnected from their predecessors? Are sentences artificially connected by a transition word hiding a progression problem? Rewrite these sentences to restore normal progression. Identify the sentences that create the wrong expectation or no expectation. They are often the very sentences that break natural sentence progression.

8

Create Reading Momentum

Cognitive neuroimaging

Michael works in a cognitive neuroscience laboratory. He explores the brain with functional MRI, and endeavours to understand what happens in our working memory. I ask him what happens when we read. Michael, an extremely well-organized man, retrieves from his computer two papers from Peter Hagoort: "Integration of word meaning and world knowledge in language comprehension",[a] and "How the brain solves the binding problem for language: a neurocomputational model of syntactic processing".[b]

Somewhat intimidated by the titles, I ask if he could explain simply what happens when we read. Still facing his Macintosh PowerBook, he quickly thinks and asks, "Do you use Spotlight?" I reply, "Of course." Any

(*Continued*)

---

[a] Hagoort P, Hald L, Bastiaansen M, and Petersson KM, "Integration of word meaning and world knowledge in language comprehension", *Science* **304**(4):438–441, 2004.

[b] Hagoort P, "How the brain solves the binding problem for language: a neurocomputational model of syntactic processing", *Neuroimage* **20**:18–29, 2003.

(Continued)

Macintosh owner with the latest operating system is familiar with the search function of Spotlight, the little white magnifying glass inside a blue spot located in the top right corner of the Mac menu bar. It is blindingly fast. "Look here," he says. I get closer to his screen. "As I type each letter in Hagoort's name, the search engine immediately updates the search results. H, then HA, then HAG. Notice how the list is now very small; one more letter, and we will have zoomed down to Hagoort's papers."

As soon as he types the letter O, the list shrinks down to a few items, and among them are Hagoort's papers. He turns towards me as I sit back into the chair facing his desk. "You see," he says, "it looks as though the Mac tries to guess what you are looking for. Similarly, while you read, your brain is active, forever seeking where the author is going with his sentence. It analyses both syntax and meaning at the same time, going from one to the other transparently."

Reading momentum is the force compelling the reader to read in order to get closure on the expectations raised by the writer in earlier sentences. Words at the beginning of a sentence can have the most powerful effect on the reader. Take a word like *although*. It immediately puts the reader on the alert. *Although* sets up the expectation that the main clause will minimise the subordinate clause. Take *because*. Placed at the beginning of a sentence, *because* announces a main clause that contains a consequence. In both cases, these conjunctions set a delay between the time the expectation is raised and the time it is fulfilled. That delay creates tension and momentum. The tension acts like a metallic spring: it pulls reading forward. In the real world, the length of a spring matters less than its strength; likewise, a sentence's

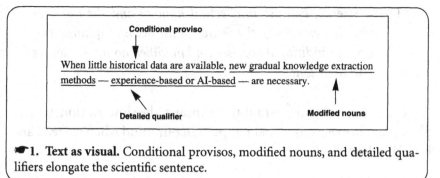

☞1. Text as visual. Conditional provisos, modified nouns, and detailed qualifiers elongate the scientific sentence.

length matters less than the tension created by the arrangement of its words. The pull of a sentence is achieved in many different ways. In this chapter, we will consider six ways.

The Text as Visual

The most powerful pull comes from visuals. Can text be a visual?

Text is usually plain. When you add style (bold, italic, underline), you make it richer. But, it can be enriched even more if you consider text as graphic. It then inherits a frame that makes it stand out, outside of paragraph text. It has its own caption and it can be annotated. In the example shown in ☞1, the main point of the visual is made in its caption: "*Conditional provisos, modified nouns, and detailed qualifiers elongate the scientific sentence.*" It is illustrated graphically. The sentence chosen is long. The graphic helps understand the caption by showing the length and by visually explaining unusual words like "*provisos*", "*qualifiers*", and "*modified nouns*".

The Subclause Hook

Placed at the beginning of a sentence, *because* creates a tension that will only be released in the main clause (the sentence's stress, i.e. the new information at the end of the sentence).

(1)   *In science, because intellectual honesty and the need for precision encourage the writer to use detailed qualifiers as well as conditional provisos and modified nouns, sentences tend to be long.*

Let us use a cinematographic technique, the slow motion, to simulate (very imperfectly) what happens in our mind when we read and how the "hook" works.

In science, (*Oh, I see, the writer mentions science to contrast it with what happens in other domains like literature or gardening.*) **because** (*The writer is giving me a reason for something here. I wonder what it is. Let's read on.*) **intellectual honesty and the need for precision** (*I know the list ends here because of the conjunction "and". So now, what do these cause?*) **encourage** (*Encourage who or encourage what?*) **the writer** (*I see, encourage the writer to do what?*) **to use** (*I'm expecting a noun now.*) **detailed qualifiers as well as** (*This looks like a list of things.*) **conditional provisos and** (*Here is the "and" conjunction again, so what follows is the last element in the list.*) **modified nouns,** (*The comma announces the start of the main clause.*) **sentences tend to be** (*Let me guess, I think I've got it, sentences tend to be complex and long, right?*) **long** (*Just as I thought. The writer had already given me the idea of length through the list, the plural, and the words "detailed" and "modified". I had no difficulty agreeing with this sentence.*)

Continuously pulled forward by questions and syntactic- or semantic-based expectations, the reader cannot stop reading. The hook works as promised.

Unlike sentence 1, in sentence 2, the subclause is located at the end of the sentence.

(2)   *In science, sentences tend to be long because intellectual honesty and the need for precision encourage the writer to use detailed qualifiers as well as conditional provisos and modified nouns.*

As a result, the stress changes. Sentence 1 emphasises length, whereas sentence 2 emphasises what creates length (the "*qualifiers*", "*provisos*", and "*modified nouns*"). If subsequent sentences elaborate on the causes of elongation, then (2), which has already prepared the ground, is better than (1). However, imagine that your readers already know that scientific sentences tend to be long. They pause briefly after reading the main clause "*sentences tend to be long*". If they are not very motivated to find out why, they will quickly scan the rest of the subclause and thus not really pay attention to why scientific sentences are long.

Instead of the usual one topic, one verb, and one stress found in simple sentences, sentence 3 has three topics ("*precision*", "*intellectual honesty*", "*sentences*"), three verbs, and three stress positions.

(3) *In science, precision requires modified keywords, and intellectual honesty demands detailed qualifiers and provisos; as a result, sentences tend to be long.*

Everything gets more emphasis. The phrase "*as a result*" is the hook that propels reading.

Sentence 4 is in the passive voice. Is it worse than (1), (2), or (3)? Let progression be your guide.

(4) *In science, sentences are usually made long by the need for precision (long modified nouns) and intellectual honesty (detailed qualifiers, caveats, and provisos).*

Three types of endings create three different reader expectations for what comes next. Sentences 1 and 3 stress length. They are preferable if the sentence that follows them elaborates on the consequences of long sentences on readers' understanding; for example, "The longer they are, the more attention they require." However, sentence 2 ends on what makes sentences long, while sentence 4 stresses the precision or intellectual honesty argument.

The Countdown

In a countdown, readers are told to expect multiple topics and/or multiple stress positions. The final closure comes when the countdown reaches zero; until then, readers remain on the alert and move forward. The countdown can be numerical, as in the next example, but it could also be a list of items reviewed in sequential order from the first to the last.

(5) *In science, two factors contribute to long sentences: precision and intellectual honesty. Adding precision to a noun through a modifier lengthens the sentence. Intellectual honesty is kept by adding detailed qualifiers and provisos to the sentence.*

The Story

The curious reader is under the spell of a story.

(6) *We were curious to find out what makes scientific sentences longer than the average book sentence. We found that the need for precision in scientific words often leads to the use of elongated modified nouns. We also discovered that, because of their intellectual honesty, scientists tend to pad their sentences with detailed qualifiers and provisos.*

The Question

A question is more powerful than a statement. Here the accent is put on the impact of long sentences on reading.

(7) *Does writing with intellectual honesty make reading difficult? It may. To be accurate, scientists tend to pad their sentences with detailed qualifiers, provisos, and packed modified nouns. Unpacking nouns, and constantly reshaping the mental image as qualifying details and provisos are added, makes reading slow and difficult.*

The Example

A word or phrase announcing an example has the same attracting power as the colon.

(8) *Intellectual honesty leads to lengthy sentences padded with detailed qualifiers such as limits or boundary conditions, and provisos such as "if" or "provided" statements. Precision has the same lengthening effect: modified nouns (nouns preceded by other nouns that modify or specify their meaning) can be two to eight words long.*

Six methods to add pull to your sentences have been presented, each with its own style and raised expectations. To decide which one is best, determine how well the sentence helps progression or sets expectations. In the end, the only thing that matters is how well your point will be understood by the reader, and how fast and pleasureable the reading experience will be. Once you start looking at words as little springs that provide pull to your sentences, your writing changes.

Variations on a theme

One musical genre has fascinated me for years: variations on a theme. Variations are mostly found in classical music. Mozart, Beethoven, and Bach have written variations on simple music themes. But, variations need not be classical in style; for example, great musicians have written variations on The Beatles' songs. Have you ever heard an opera singer sing The Beatles' song "Blackbird", a jazz musician swing on "Blackbird", or a Japanese rock band rock on "Blackbird"? They all sound different, yet the melody is never lost. The melody of your paper is its contribution. What is your style?

Read your introduction. How much pull do your sentences have? The pull of cotton thread, the pull of rubber band, or the pull of steel spring? Bring some pull back into your sentences through one of the ways presented in this chapter. A word of advice: when adding pull, think ahead, i.e. consider the next sentence and keep progression in mind.

9

Control Reading Energy Consumption

Réponse hémodynamique

The article by Peter Hagoort that Michael had given me to read was truly fascinating. What happens in our brain when, during reading, it encounters strange things such as "the car stopped at the casserole traffic light"? Something similar happened to me while reading the word "hemodynamic" in the article. Google took me to the website fr.wikipedia.org/wiki/Réponse_hémodynamique, and then things became very interesting. I discovered that when reading becomes difficult, the body sends a little more blood (i.e. glucose and oxygen) to the brain. It does not take blood from one part of the brain to send it to another part so as to keep energy consumption constant; it simply increases the flow rate. Following the trail like a bloodhound, I discovered a French article written by André Syrota, director of the life science division at the Atomic Energy

(Continued)

> *(Continued)*
>
> Commission, indicating that our brain's additional work could consume the equivalent of "147 joules per minute of thought".[a]

How tired will your readers be at the end of their reading journey? How well did you manage their time and energy? As Gopen[b] points out, reading consumes energy. Reading scientific articles consumes A LOT MORE ENERGY. Therefore, how do you reduce the reading energy bill, and how will you give your reader the assurance that plenty of energy-refuelling stations will be available along the long and winding road of your text?

The Energy Bill

Let, E_T, be the total energy required by the brain to process one sentence. E_T is the sum of two elements: the syntactic energy E_{SYN} spent on analysing sentence structure, and the semantic energy E_{SEM} spent on connecting the sentence to the others that came before it and on making sense of the sentence based on the meaning of its words.

$$E_T = E_{SYN} + E_{SEM}.$$

E_T is quasi-finite and is allocated by the brain to the reading task. Similar to our lungs, which give us the oxygen we need one breath at a time, the brain has enough energy to read one sentence at a time. E_T is not completely finite, but it cannot increase beyond a certain limit fixed by physiological mechanisms: to increase the blood flow rate takes a few seconds, and the size of the blood vessels in the brain (although extensible) is limited. Therefore, we will assume that, once allocated, E_T is constant. This means that if E_{SYN} becomes large, it will

[a] http://histsciences.univ-paris1.fr/i-corpus-evenement/FabriquedelaPensee/affiche-III-8.php

[b] Gopen GD, *Expectations: Teaching Writing from the Reader's Perspective*, Pearson Longman, p. 10, 2004.

be at the expense of E_{SEM}: the more energy is spent on the analysis of the syntax of a sentence, the less energy will be left to understand its meaning. Gopen[c] considers these two energies to be in a "zero-sum relationship".

You cannot increase E_T, the total reading energy, because the reader controls it. You can, however, make sure that $E_{SYN} + E_{SEM} < E_T$ at all times by minimising both the syntactic and semantic energies required to read.

What would consume excessive syntactic energy, E_{SYN}?

1. Anything ambiguous or unclear — a pronoun referring to an unclear noun, a convoluted modified noun, an ambiguous preposition.
2. Spelling or light grammar mistakes, such as a missing *the*, the preposition *in* instead of *on*, the verb *adopt* instead of *adapt*.
3. Incomplete sentences, i.e. missing verb.
4. Anything taxing on the memory — long sentences (usually written in the passive voice) with long modified words, formulas, multiple caveats, provisos, long qualifiers, sentences with deeply nested subordinates.
5. Grammatical structures from a foreign language applied to English without modification.
6. Missing or erroneous punctuation.

What would consume little syntactic energy, E_{SYN}?

1. Small sentences with simple syntax: subject, verb, object.

   New ideas disrupt the logical flow of sentences.

2. Sentences with a predictable pattern established with words such as *although, because, however,* or *the more … the less.*

---

[c] Gopen GD, *op. cit.*, p. 11, 2004.

The more energy is spent to analyse the syntax of a sentence, the less energy is left to understand what the sentence means.

3. Sentences with subject close to verb, and verb close to object.

Motivation allocates the total energy E_T to the reading task.

4. Sentences with good punctuation.

The reader has three choices: give up reading, read the same sentence again, or read what comes next.

What would consume great semantic energy, E_{SEM}?

1. Unknown words, acronyms, and abbreviations.
2. Absence of context to derive meaning.
3. Lack of prior knowledge to understand or to aid understanding.
4. Lack of examples or visuals to make the concept clear.
5. Overly detailed or incomplete visuals.
6. Reader forgetful of content previously read.
7. Reader in disagreement with statement, method, or result.
8. Very abstract sentences (formulas).
9. Sentences out of sync with reader expectations.

What would consume little semantic energy, E_{SEM}?

1. A sentence with a well-established context.

Total reading energy for a given sentence, E_T, is the sum of two elements: the syntactic energy E_{SYN} spent on analysing its syntax, and the semantic energy E_{SEM} spent on making sense of the just analysed sentence.

$$E_T = E_{SYN} + E_{SEM}.$$

2. A reader familiar with the topic or the idea.

The songbird flew back to the nest to sit on three little eggs; two of them its own, the third one from a cuckoo.

3. A sentence that explains the previous sentence.

   Therefore, if E_{SYN} becomes large, it will be at the expense of E_{SEM}. The more energy is spent to analyse the syntax of a sentence, the less energy will be left to understand the meaning of the sentence.

4. A sentence that prepares the grounds (through progression or setting of context).

   Subclauses that pull reading forward often follow a predictable pattern: they start with a preposition such as although, because, however, *or* if.

5. Short sentences (with known vocabulary).

   It does not. The reader is surprised.

What would get the reader into trouble?

Energy shortages occur when $E_{SYN} + E_{SEM} > E_T$.

1. E_{SYN} is unexpectedly large. As a result, what remains of E_{SEM} is insufficient to extract the complete meaning of the sentence.
2. E_{SYN} is normal; but a new word, acronym, abbreviation, apparent contradiction, or concept requires additional brain effort (saturated memory, or failure to find associative link with known data). The reader runs out of E_{SEM}. The semantic energy gas tank is empty before the sentence is fully understood.

When this happens, the reader can make one of three choices: give up reading; read the same sentence again; or read what comes next, hoping to understand later.

Giving up reading is tragic. It is a consequence of repetitive and successive breakdowns in understanding. Usually, the reader will continue to read, hoping to understand later. Sometimes, they do understand; but more often, the text becomes more and more obscure, and the reader finally gives up reading.

Rereading takes place if the reader's motivation is high. The reader is determined to understand, or much understanding is expected from the difficult sentence. The rereading that occurs because of a difficult syntax consumes no syntactic energy: the sentence syntax is now familiar, and the reader can spend all of his or her energy on understanding the text.

$$E_{SYN} = 0 \quad \text{and therefore} \quad E_T = E_{SEM}.$$

The metaphor of reading as consuming brain energy is in line with what science observes. The brain that is hard at work consumes more energy.

The Role of Motivation

Attention is precious. One should not waste it. It directs the activity of the brain. Attention is a thought traffic controller. If attention wanes, our train of thoughts could derail or be redirected to another set of rails. Yet, for all its importance, attention is governed by a powerful ruler: motivation. Motivation determines the total energy level ΣE_T allocated to the reading task.

Consider reading as a system with inputs and outputs, as shown in ☛1. Motivation is one of the five critical inputs to the system. The reader's initial need or interest feeds it. The fun of gaining knowledge (feedback loop) keeps it high.

Gaining motivation is internal to the reading process, a result of it. Motivation gains occur when expectations are exceeded or when goals

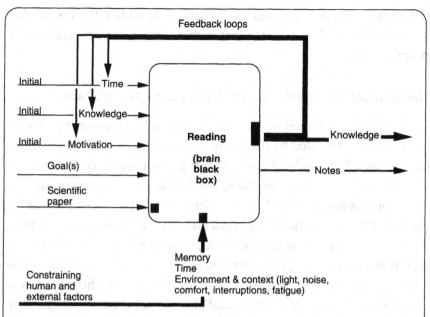

☛1. Reading: a system perspective. Reading, considered as an open system, has five inputs and two main outputs. Prior to reading, each input has an initial value. This value will change over time because the outputs influence the inputs. For example, the more knowledge you get from a paper, the more knowledge you put back to facilitate further understanding. External factors also influence the reading process. They either lubricate the process or create friction and inefficiencies. They indirectly impact the pace of absorption of knowledge and therefore motivation, a critical input to the system. If reading was a transistor motivation would be its base current that either shuts down or promotes the reading activity.

are met quickly. Losing motivation is both internal and external to the reading process: internal when expectations are not met (syntax is too obscure or initial knowledge is insufficient), and external when alternatives to reading become more attractive or when the reader is tired.

Punctuation: a Refuelling Station

The full stop: a period to refuel

When the full stop (period) arrives, the reader pauses and refills his or her energy tank before reading the next sentence. It gives the reader a chance to conclude, absorb, consolidate the knowledge just acquired, and anticipate what comes next (from expectations or progression).

The semicolon: a fuel stop for topping a half-full tank

Surprisingly, searching for a semicolon through a scientific paper will often be rewarded by the infamous beep that says, "None found, can I search for anything else?" Periods, colons, and commas seem to be the only punctuation marks used by scientists. Semicolons are close cousins to the period. They always stand at a place of semantic closure. Like the period, they end and start a main clause. Unlike the period, their role is to unite, join, or relate; while the role of the period is to separate. The main clauses on each side of a semicolon are often compared, contrasted, or opposed. Often, the first clause in the sentence makes a point, and the clauses after the semicolon refine, detail, or complete the point. Semicolons are found where conjunctive adverbs such as *consequently, however, therefore, thus,* or *nonetheless* are used.

> *The calculated data and the observed data were closely related; however, the observed data lagged behind when concentration dropped.*

Scientists, by nature logical, should be fond of semicolons not only to strengthen their arguments, but also to make their text less ambiguous and to carry the context forward at little cost. The reason for the latter is simple: the two clauses joined by a semicolon are closely related semantically, much more so than two sentences separated by a period. Therefore, since the context does not vary within the sentence, reading is faster and easier.

A semicolon has more than one use. When a sentence needs to be long to keep together a list of sentences, the semicolon does the job magnificently.

Information with visual impact requires creativity, graphic skill, and time. Because most of these are in short supply, software producers provide creativity, skill, and time-saving tools: statistical packages that crank out tables, graphs, and cheesy charts in a few mouse clicks; digital cameras that, in one click, capture poorly lit photos of experimental setups replete with noodle wires (I suppose the more awful they look, the more authentic they are); and screen capture programs that effortlessly lasso and shrink your workstation screen to make it fit in your paper.

The :!? fuel stops and the comma

Other punctuation marks also provide an opportunity to refuel: the colon, the question mark, and the unscientific exclamation mark (I wonder if Archimedes would have damaged his reputation as a scientist had he ended his "Eureka" statement with an exclamation mark). The colon introduces, explains, elaborates, recaps, and lists. Unlike the semicolon, it can be followed by a phrase that lacks a verb. Like the semicolon, it is preceded by a whole main clause (not a truncated one, as in the next example).

And the results are:

In a correct sentence, the main clause is not truncated.

And the results are the following:

Colons are much liked by readers: they announce clarification or detail. Colons are also the allies of writers. They help to introduce justification after a statement.

Commas help to disambiguate meaning, pause for effect, or mark the start and end of clauses. But, for all their qualities, there is one that commas cannot claim: semantic closure. Readers cannot stop at a comma and decide that the rest of the sentence can be understood without reading further.

In this chapter, you have been given many tools to reduce the reading energy bill of the reader. Imagine your writing as a piece of cloth, and the brain of the reader as an iron. If your writing has the smoothness of silk, the iron can be set at the lowest temperature setting. If it has the roughness of overdry cotton, not only will the iron have to be set at the highest temperature setting, but you will also put the reader under pressure and demand steam to iron out the ugly creases in your prose. It is a zero-sum game. Either you spend time and energy, or the reader does.

Ask a reader to read your paper and to highlight in red the sentences not clearly understood, and in yellow the sentences that slowed down reading because of a difficult syntax. Then, correct accordingly.

Part II

Paper Structure and Purpose

Each stage of the construction of a house contributes to its overall quality. Similarly, each part of an article contributes to the quality of the whole, from the abstract (the architectural blueprint) and the structure (the foundations) to the introduction (the flight of steps and the landing in front of your main door), the visuals (the light-providing windows), and finally the conclusions (the handing out of the key to knowledge). The art of construction is acquired through a long apprenticeship. You may be attracted by the time-saving expedient prefab (even its name indicates that it is a shortcut), or by the imitation of other constructions of uncertain architectural quality. Beware of shortcuts. A thorough analysis of the different parts of a hastily assembled paper often reveals major cracks and faults: the shapeless structure is like a pair of baggy jeans that fit just about any frame, while the graphics and other visuals have a mouse- and mass-produced look and feel.

To construct a satisfactory set of parts, one must understand the role played by each for the reader and the writer; and to assess their quality, one must establish evaluation criteria. The next chapters fulfil these objectives. Numerous examples are given for analysis and to help distinguish good writing from bad writing.

First impression

Today, as the city's bowels demonstrate their usual constipation, the pouring rain adds a somewhat slimy aspect to the slow procession of traffic. Professor Leontief does not like arriving late at the lab. He hangs his dripping umbrella over the edge of his desk, at its designated spot above the trashcan, and he gently awakens his sleepy computer with some soothing words: "Come on, you hunk of metal and silicon oxide, wake up."

He checks his electronic mail. The third e-mail is from a scientific journal which he helps out as a reviewer. "Dear Professor Leontief, last month you kindly accepted to review the ...." He need not read any further. He looks at his calendar, and then feels the cold chill of panic run up his spine when he realises that the deadline is only 2 days away. He hasn't even started. So much to do with so little time! Yet, he cannot postpone his response. Being a resourceful man, he makes a couple of telephone calls and reorganises his work schedule so as to free up an immediately available 2-hour slot.

He pours himself a large mug of coffee, and extracts the article from the pile of documents pending attention. He goes straight to the reference section on the last page to check if his own articles are mentioned. He grins with pleasure. As he counts the pages, he looks at the text density. It shouldn't take too long. He smiles again. He then returns to the first page to read the abstract. Once read, he flips the pages forward slowly, taking the time to analyse a few visuals, and then moves to the conclusions, reading them with great care.

(Continued)

> (*Continued*)
>
> He stretches his shoulders and takes a glance at his watch. Twenty minutes have gone by since he started reading. By now, he has built a first and strong impression. Even though the article is of moderate length, it is too long for the depth of the proposed contribution. A letter would have been a more appropriate format than a full-fledged paper. Poor researcher. He will have to say this, using diplomatic skills so as not to be discouraging, for he knows the hopes and expectations that all writers share. What a shame, he thinks. Had he accepted the paper, his citation count would have increased. Now the hard work of thorough analysis lies ahead. He picks up his coffee mug and takes a large gulp.

The first impression of a paper is formed after a partial reading. During the first 20 minutes or so, a reviewer does not have time to read the whole paper, in particular the methodology and the results/discussion sections. I have therefore decided to cover in part II only those parts of a paper that are read during the rapid time in which the first impression is formed. This decision was also based on comments from scientists who have published many papers. They stated that the methodology and results sections of their paper were the easiest and fastest to write, but it was the other parts that were difficult and took time: the abstract, introduction, and conclusions. As for the title, structure, and visuals, they recognised that they had underestimated the key role these parts play in creating the first impression.

The impact of the quality of these parts goes beyond creating a favourable first impression for the reviewer and reader. Improved readability and more clearly expressed scientific contribution will generate more feedback from the scientific community. The difference between making ripples or making waves will then be a matter of scientific excellence — a topic I leave in your good and capable hands!

10

Title: The Face of Your Paper

When I think about the title of a paper, quite naturally, the metaphor of a face comes to mind. So many features of a title resemble those of a face. *First among them is what people call the "first impression": it is your face people look at to get a first impression of you.* Likewise, a title contains the first words the readers will see. It will give them a first impression of how well your paper meets their needs and whether or not it is worth reading. *Your face sets expectations as to the type of person you are.* Your title will also reveal what kind of paper you have written, its breadth, and its depth. *Your face is unique and memorable. It is found on your passport and various official documents.* Your unique title will be found in references and databases. *What makes your face unique is the way its features are assembled harmoniously.* What makes your title unique is the way its keywords are assembled to differentiate your work from the work of others.

When I was 12 years old, I stumbled upon a strange book in my local library. It was about morphopsychology — the study of people's characters as revealed by the shape of their faces. I do not remember much about it today, but I do remember it was fun. Discovering a paper from its title should also be fun. In the following dialogue, imagine yourself as the scientist answering the questions. How would you answer?

Six Titles to Learn About Titles

Author: Greetings, Mr Scientist. I'd like to introduce a series of six titles and ask you one or two questions about each one. These titles may be in areas you are not familiar with, but I'm sure you'll do fine. Are you ready?

Scientist: By all means, go ahead!

Author: All right then. Here is the first title.

"Gas-assisted powder injection moulding (GAPIM)" [a]

Based on its title, is this paper specific or general?

Scientist: Hmm, you are right, I know nothing about powder injection moulding. The title seems halfway between being specific and being general. "Powder injection moulding" by itself would be general, maybe a review paper. But, this title is a little more specific. It says "Gas-assisted", which seems to indicate that there are other ways to do powder injection moulding.

Author: You are right. GAPIM is used to make hollow ceramic parts. People in that field would be quite familiar with powder injection moulding and its PIM acronym. What would have made the title more specific?

Scientist: The author could have mentioned a new specific application for GAPIM.

---

[a] Li Q, William K, Pinwill IE, Choy CM, and Zhang S, "Gas-assisted powder injection moulding (GAPIM)", *International Conference on Materials for Advanced Technologies (ICMAT 2001), Symposium C: Novel and Advanced Ceramic Materials*, Singapore, 2001.

Author: Good. How do you feel about the use of the GAPIM acronym in the title?

Scientist: I am not sure it is necessary. I have seen acronyms in titles before, but they were used to launch a name for a new system, a new tool, or a new database. The acronym was usually more memorable than the long modified name it replaced. Unless this is the first article ever published on this technology, in my opinion, it is not necessary to use an acronym.

Author: Thank you. How about this second title: general or specific?

"Energy-efficient data gathering in large wireless sensor networks"[b]

Scientist: This title is very specific. It mentions the domain "wireless sensor networks", and makes it even more specific by adding the adjective "large". The contribution seems clear: "energy-efficient". This adjective hints that data gathering is not energy-efficient when the network is large. I know nothing in this domain either, but it seems to make sense.

Author: You are perfectly entitled to logically infer that from the title. Actually, all readers generate hypotheses and expectations from titles. How about these two titles: are they both claiming the same thing?

"Highly efficient waveguide grating couplers using silicon-on-insulator"

"Silicon-on-insulator for high-output waveguide grating couplers"

Scientist: Well, I suppose the first paper is mostly about waveguide grating couplers, and the second about Silicon-on-insulator. What comes first in the title, usually the author's contribution, is the most important information.

---

[b] Lu KZ, Huang LS, Wan YY, and Xu HL, "Energy-efficient data gathering in large wireless sensor networks", *Second International Conference on Embedded Software and Systems (ICESS'05)*, Xi'an, China, pp. 327–331, 2005.

Author: Bravo! You are doing fine. Now, look at the following two titles. Besides the use of an em dash or a colon to introduce the benefit of web services, are these two titles equivalent?

"*Web services — an enabling technology for trading partners community virtual integration*"[c]

"*Web services: integrating virtual communities of trading partners*"

Scientist: Um ... this is a difficult one. The long five-word modified noun in the first title is difficult to read, yet I am attracted by the catchy term "enabling technology". The second title does not have the problems of the first. It is shorter, more dynamic, and purposeful. But, is it necessary to put a colon after "web services"? The second part of the title does not really explain or illustrate web services. Could the title be changed to "Integrating virtual communities of trading partners through web services"? In this way, what is new comes at the beginning of the title. I don't think that web services are really new.

Author: The title could be changed to what you propose. You are right; the second title is more dynamic. The use of the verbal form "integrating" makes it so. You are doing very well. Only two more titles.

"*Vapor pressure assisted void growth and cracking of polymeric films and interfaces*"[d]

Scientist: Vapor with an "o". It is for an American journal, isn't it? If it had been for a British paper, they would have written "vapour". One has to be careful with keyword spelling nowadays, even if the scientific search engines are getting better. Fortunately, the title contains many keywords, so I would have found it. If I may, I would like to add something.

---

[c] Lee SP, Lee HB, and Lee EW, "Web services — an enabling technology for trading partners community virtual integration", *Fourth International Conference on Electronic Business (ICEB 2004)*, Beijing, China, pp. 727–731, 2004.

[d] Cheng L and Guo TF, "Vapor pressure assisted void growth and cracking of polymeric films and interfaces", *Interface Sci* 11(3):277–290, 2003.

Author: Go ahead.

Scientist: This title contains two "and" conjunctions, which create ambiguity. I do not know if there are two contributions in this paper ("Vapor pressure assisted void growth AND cracking of polymeric films and interfaces") or only one ("Vapor pressure assisted void growth and cracking of polymeric films and interfaces"). The second "and" is just as ambiguous: does the adjective "polymeric" apply to films and interfaces, or only to films? I am sure an expert would not find the title ambiguous, but nonexperts like myself would.

Author: Excellent observation. Titles have to be clear to all, experts and nonexperts. Besides *and* and *or*, other prepositions can also be quite ambiguous in titles. For example, the preposition *with* could mean *together with* as in "coffee with milk", or it could mean *using* as in "to move the ground with a shovel".

The time has come for our last title. It is somewhat tricky. Can you identify the author's contribution?

> "A new approach to blind multiuser detection based on inter-symbol correlation"

Scientist: Other researchers are already doing research in this field, and the author is following the pack with a new approach. Personally, I don't like the word "approach": it is vague, whereas the words it replaces are more specific. I would use "method", "technique", "system", "algorithm", or "technology" instead. I also don't like titles that start with "a new" something. In my opinion, it never takes long before someone else develops a newer technique. Furthermore, "new" does not indicate what is new or what makes it new. As for the contribution of this paper, I must say I am at a loss. The intersymbol correlation could be new, but if that is the case, why is it at the back of the title? It should be at the front. "Intersymbol correlation for blind multiuser detection" is clear. Or (and I suspect this is the case), intersymbol correlation is not new, but the author has modified the method. That would explain the use of "based on". In that case, why doesn't he tell us either the benefit of *the modified method or the method for this modification? It would be more informative and more compelling.*

Author: You are quite good at this. Thank you so much for assisting me in this dialogue.

Scientist: Not at all!

Less time than you think

Have you ever considered how readers access your title and read it? I do not mean to be a killjoy, but your title is not read: it is scanned, within 2 seconds at the most. Appalling, isn't it! You spend 9 months researching and 2 full weeks writing the paper, but readers will decide whether to read your paper or not in a second or two! If you do not generate interest in that extremely short time, forget about being read, forget about citations, and forget about making an impact on science.

Your title is usually one of many titles retrieved by the search engine and presented in list form. It may be anywhere on the list. Reading a list is not like reading text in the context of a paragraph. Each item on the list stands alone, without context. The only thing you know is that every title on the list contains one or several of your search keywords. What does one have time to do in 2 seconds? Word spotting, mostly. You will pay more attention to the words that surround the search keywords. The rest of the title will be glided over. A short title is better than a long one, but an easily understood long title is better than a short title whose nouns need unpacking to be understood.

You need to impress. To do that, you have less than 2 seconds of the reader's time!

Six Techniques for Improving Titles

Placement of contribution upfront in a title

In a full sentence (containing a verb), new information usually appears at the end (stress position) and old information at the beginning (topic position). In a verbless title, however, the situation is reversed: new information (i.e. the contribution) appears at the beginning; and the known, less specific information, at the end.

Addition of verbal forms

A phrase without a verb lacks energy. The gerundive and infinitive verbal forms add energy to a title.

*"Data learning: **understanding** biological data"*[e]

*"Nonlinear finite element simulation **to elucidate** the efficacy of slit arteriotomy for end-to-side arterial anastomosis in microsurgery"*[f]

Adjectives and numbers to describe the strong point of a contribution

Besides specific keywords, adjectives and adverbs are often used to describe the key aspect of a contribution — *fast, highly efficient,* or *robust* (avoid *new* or *novel*). Since adjectives are subjective, replacing them with something more specific is always better. A "20 Ghz thyristor" is clearer than a "fast thyristor"; and while in 20 years "fast" will make a liar out of you, "20 Ghz" will not.

---

[e] Brusic V, Wilkins JS, Stanyon CA, and Zeleznikow J, "Data learning: understanding biological data", in Merrill G and Pathak DK (eds.), *Knowledge Sharing Across Biological and Medical Knowledge-Based Systems: Papers from the 1998 AAAI Workshop*, AAAI Press, Menlo Park,CA, pp. 12–19, 1998.

[f] Reprinted from Gu H, Chua A, Tan BK, and Hung KC, "Nonlinear finite element simulation to elucidate the efficacy of slit arteriotomy for end-to-side arterial anastomosis in microsurgery", *J Biomech* **39**: 435–443, 2006 (with permission from Elsevier)

Clear and specific keywords

The specificity of a paper is proportional to the number of specific keywords in its title. Beware of keywords buried in long modified nouns, because their clarity is inversely proportional to the length of the noun. Modified nouns are slightly more concise, but often at the expense of clarity.

"Transient model for kinetic analysis of electric stimulus-responsive hydrogels" **(unclear)**

"Transient model for kinetic analysis of hydrogels responsive to electric stimulus" **(clear)**

Sometimes, keywords change their spelling when embedded inside a modified noun. *Segmentation* may become *segmented* or *segmenting.* If the most frequent word used for retrieval is *segmentation,* your title may not be found; or if it is, it may not be listed among the first 10 titles retrieved.

Smart choice of keyword coverage

Even when published, an article will have little impact if it is not found. Readers find new articles through online keyword searches. Choosing effective keywords is vital. If you pick your keywords from recent or often-cited titles close to your contribution, then searches that retrieve these articles will also retrieve yours and so the chances of it being read will increase.

When two different keywords with the same meaning appear with the same frequency in titles, choose one for the title and the other for the abstract. That way, the search engines will find your paper, regardless of the keyword used for the search.

Keywords are divided into three categories (☛1). General keywords (*simulation, model, chemical, image recognition, wireless*

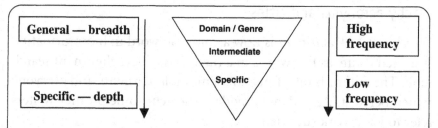

☛**1. Keyword depth and breadth.** Specialised keywords are at the pointed lower end of the inverted triangle. General keywords are at the broad top end of the triangle. The general-to-specific scale correlates with the frequency of use of a scientific keyword. Depth and breadth of a keyword are not intrinsic qualities, but rather depend on the frequency of use of these words in the journal that publishes the paper. For example, the reader of *Science* may consider "nanopattern" very specific, yet the reader of the *Journal of Advanced Materials* will find it quite generic. The reader's knowledge also influences the perception of keyword levels: the less knowledgeable the reader is, the more the general keywords will seem specific, and vice versa.

network) are useful to describe the domain or the type of your work/paper, but they have very little differentiating power precisely because they frequently appear in titles. They do not help to place your title at the top of the reader's list. Intermediate keywords are better at differentiating. They are usually associated with methods common to several fields of research (*fast Fourier transform, clustering, microarray*) or to large subdomains (*fingerprint recognition*). But, for maximum differentiation, specific keywords are unbeatable (*hypersurface, hop-count localisation, nonalternative spliced genes*). For a given journal, or for domain experts, the category of a keyword is well defined. It changes from journal to journal, or from experts to nonexperts.

Make sure your title has keywords at more than one level of the triangle. If too specific, your title will only be found by a handful of experts in your field; it will also discourage readers with a sizeable knowledge gap. If too general, your title will not be found by experts. The keyword choice decision is yours. Be wise.

Catchy acronyms and titles

The BLAST acronym is now a common word in bioinformatics. It started its life as five words in a title: "Basic local alignment search tool". The author built a fun and memorable acronym, and everyone remembered it. Acronyms provide a shortcut to help other writers refer to your work succinctly.

> *"VISOR: learning VIsual Schemas in neural networks for Object Recognition and scene analysis"* [g]

The title above is that of the doctoral thesis of Wee Kheng Leow. Other researchers mentioning his work could, for example, write "in the VISOR system [45]". The acronym provides a convenient way for others to refer to his work. Notice that both BLAST and VISOR are memorable. Acronyms like GLPOGN are doomed to fail.

Here is a catchy and intriguing title.

> *"The diner-waiter pattern in distributed control"* [h]

"Distributed control" is not usually associated with the interaction between a restaurant waiter and a customer. What the title gains in interest, however, it loses in retrieveability: it only has one general domain keyword (*"distributed control"*), and researchers in this domain are unlikely to even think of *"diner-waiter"* as a search keyword. But, if the diner–waiter pattern represents a significant scientific contribution, it will be presented at a conference or be accepted in a tier-one journal. Scientists will then take note of it, refer to it, and the rest is history. Therefore, if you conduct cutting-edge research, do not let specific keywords restrain your choice of title words.

---

[g] Leow WK, "VISOR: learning visual schemas in neural networks for object recognition and scene analysis", PhD dissertation, Technical Report AI-94-219, 1994.

[h] He H and Aendenroomer A, "Diner-waiter pattern in distributed control", *Proceedings of 2nd International Conference on Industrial Informatics (INDIN'04)*, Vol. 2, Berlin, Germany, pp. 293–297, 2004.

A pioneering article can also be retrieved through the author's name, citations, references, or abstract keywords. Be aware that some search engines give more importance to words in a title than to words in an abstract.

The question makes a mighty hook.

"Software acceleration using programmable logic: is it worth the effort?" [i]

Beware of making a title catchy by using an expression that does not make sense across different cultures. Would you understand these titles?

"The inflammatory macrophage: a story of Jekyll and Hyde" [j]

"The abc's (and xyz's) of peptide sequencing" [k]

> **The benefits of being first**
>
> If you are a pioneer in your field, the choice of words is entirely yours. Since you are the first to write in this field, you need not worry about titles that may have already been used. Think about it. Imagine being the first to write about dialogue in speech recognition. Finding a title is easy. Now, imagine you are the 856th writing a paper in this crowded field. You have to be much more specific to differentiate your title from the others. As a result, you might have to settle for a long specific title like "Semantic-based model for multiphase parsing of spontaneous speech in dialogue systems".

---

[i] Edwards M, "Software acceleration using programmable logic: is it worth the effort?" *Proceedings of the 5th International Workshop on Hardware/Software Codesign,* Braunschweig, Germany, pp. 135–139, 1997.

[j] Duffield JS, "The inflammatory macrophage: a story of Jekyll and Hyde", *Clin Sci (Lond)* **104**(1):27–38, 2003.

[k] Steen H and Mann M, "The abc's (and xyz's) of peptide sequencing", *Nat Rev Mol Cell Biol* **5**:699–711, 2004.

Purpose and Qualities of Titles

Purpose of the title for the reader

1. It helps the reader decide whether the paper is worth reading further.
2. It gives the reader a first idea of the contribution: a new method, chemical, reaction, application, preparation, compound, mechanism, process, algorithm, or system.
3. It provides clues on the type of paper (review paper or introductory paper), its specificity (narrow or broad), its theoretical level, and its nature (simulation or experimental). By the same means, it helps the reader assess the knowledge depth required to benefit from the paper.

Purpose of the title for the writer

1. It allows the writer to place enough keywords for search engines to find the title.
2. It catches the attention of the reader.
3. It states the contribution in a concise manner.
4. It differentiates the title from other titles.

Qualities of a title

Now that you know the purpose of a title, you are in a better position to write one that serves both you and the reader. Once written, how will you evaluate the title quality? Here are a few adjectives to help you.

A title is **UNIQUE.** It differentiates your title from all others (present or future).

A title is **LASTING.** Try not to use *new* in it. A title may outlive you. Ask Darwin!

A title is **CONCISE.** Some keywords are overly detailed. Remove the details if your title is unique without them.

A title is **CLEAR.** Avoid long modified nouns (a major source of misunderstanding and imprecision).

A title is **EASY TO FIND.** Its keywords are carefully chosen.

A title is **HONEST and REPRESENTATIVE** of the contribution and the paper. It sets the expectations and answers them.

A title is as **CATCHY** as can be. Remember, you only have one chance and 2 seconds to interest the reader.

A Title to Test Your Skills

Let us test our understanding of these qualities on this title:

"Hydrophobic property of sol-gel hard coatings"[1]

---

[1] Wu LYL, Soutar AM, and Zeng XT, "Hydrophobic property of sol-gel hard coatings", Paper ID: 34-TCR-A500, *Proceedings of the 2nd International Conference on Technological Advances of Thin Films and Surface Coatings* (Thin Films 2004), Singapore, pp. 13–17, 2004.

This title is short and interesting. The reader expects an article reviewing one property of various sol-gel hard coatings. Now, imagine that the article is really about ways to increase the hydrophobicity. Would the following title be better?

"Increasing hydrophobicity of sol-gel hard coatings by chemical and morphological modifications" [m]

Has the quality of the title improved? It is more **representative** of the contribution of the paper. It is **honest** because it does not claim that it will reveal all about the hydrophobic property of sol-gel hard coatings. It is **easier to find** because it adds keywords. Moreover, it is **clearer** because it mentions how this increase in hydrophobicity is achieved. Although it has lost conciseness because it is longer, it has gained in appeal because it uses a verbal form (*"increasing"*).

This title is quite catchy:

"Increasing hydrophobicity of sol-gel hard coatings by mimicking the lotus leaf morphology"

"Lotus leaf" is unexpected. The title may attract scientists outside the domain of manufacturing technology, or journalists writing for more widely distributed science magazines. However, some keywords describing the methodology have been lost (*"chemical and morphological modifications"*).

A good title attracts the reader and enhances your chances of being cited. It is fair to say that readers familiar with a research field search by keyword less often than they search by author or citation. The latter search is quicker and more fruitful. But first, you must

---

[m] Wu LYL, Soutar AM, and Zeng XT, "Increasing hydrophobicity of sol-gel hard coatings by chemical and morphological modifications", *Surface and Coatings Technology* **198**(1–3):420–424, 2005.

become an author whose name is sought — this starts with good research and good titles.

Catchy title . . . but how?

Here are seven proven ways:

(1) Adjectives are attractive.

(2) Some keywords carry the passion of the time. Encountering them in titles excites the reader who is keen to keep up to date with the latest happenings in science.

(3) Verbal forms (gerundive and infinitive) are more active and potent than strings of nouns connected by prepositions.

(4) A shorter title is more attractive than a long one, and a general title is more attractive than a specific one.

(5) Words that announce the unexpected, the surprising, or the refutation of something well established all fuel the curiosity of the reader.

(6) Unusual words that belong to a different lexical field intrigue the reader.

(7) Questions are great, but are often reserved for the few who have reached professorship or Nobel Prize status.

To make a title catchy, there is only one rule: catchy, yes; dishonest, no.

What do you think of your title? Does it have enough of the qualities mentioned here? Is your contribution featured at the head of your title? It is time to have a closer look.

11

Abstract: The Heart of Your Paper

The heart plays an essential role in the human body. Similarly, the essence of an article is its abstract. It goes to the core. *The heart has four chambers.* The abstract is also composed of four easily identifiable parts.

> **Visuals in abstracts?**
>
> Never say never! I used to think that abstracts had no visuals, but it looks as though I was mistaken. The tables of contents of some journals (e.g. *Advanced Materials, Journal of the American Chemical Society*) now include a key visual alongside an abridged abstract. Is this a preview of the shape of things to come for all journals? I believe it is. A good figure far exceeds plain text in illustrating and explaining a contribution efficiently and concisely. Therefore, take note and prepare yourself.

The abstract dissected in this chapter is at the crossroads between surgery and computer science. It comes from a paper on slit arteriotomy. The easiest way to explain it is to visualise anastomosis — the surgical connection of two tubes (here, arteries). Normally, the surgeon cuts an elliptic hole (with removal of material) in the recipient artery and then stitches the donor artery over the hole. In this case, however, only a slit is cut in the side of the recipient artery before the donor artery is stitched over it. Consequently, there is no need to remove any material. Does slit arteriotomy work as well as hole arteriotomy?

Surgeons are (with good reason) very conservative: if a procedure (hole arteriotomy) works, why replace it with a new one (slit arteriotomy), even if initial statistics convincingly establish that the new technique is equivalent to the conventional one? To establish the safety and efficacy of the new technique, the surgeon who invented it asked for the help of computer-modelling scientists. The technique was modelled, and a paper was born. Its title was this:

> *"Nonlinear finite element simulation to elucidate the effi-cacy of slit arteriotomy for end-to-side arterial anastomosis in microsurgery"* [a]

The title is composed of two parts: contribution and background. If you were to put a dividing bar | between these two parts, where would you put it? The answer will come later, after you have read the abstract. Note that the words in bold are common to both the abstract and the title.

> "[61 words] **The slit arteriotomy for end-to-side arterial microanastomosis** *is a technique used to revascularize free flaps in reconstructive surgery. Does a slit open to a width sufficient for blood supply? How is the slit opening affected by factors such as arterial wall thickness and material stiffness? To answer these*

[a] Reprinted from Gu H, Chua A, Tan BK, and Hung KC, "Nonlinear finite element simulation to elucidate the efficacy of slit arteriotomy for end-to-side arterial anastomosis in microsurgery", *J Biomech* **39**:435–443, 2006 (with permission from Elsevier).

*questions we propose **a nonlinear finite element** procedure to simulate the operation. [10 words] Through modeling the arteries using hyperelastic shell elements, our **simulation** [112 words] reveals that the slit opens to a width even larger than the original diameter of the donor artery, allowing sufficient blood supply. It also identifies two factors that explain the opening of the slit: blood pressure which is predominant in most cases, and the forces applied to the slit by the donor artery. During simulation, when we increase the donor artery thickness and stiffness, it is found that the contribution of blood pressure to the slit opening decreases while that of the forces applied by the donor artery increases. This result indicates that sometimes the forces by the donor artery can play an even more significant role than the blood pressure factor. [28 words] Our simulation **elucidates the efficacy** of the slit arteriotomy. It improves our understanding of the interplay between blood pressure and donor vessel factors in keeping the slit open. [Total: 211 words]"*[b]

The Four Parts of an Abstract

Each of the four parts in the abstract above (separated by the word count) answers key questions that the reader has.

Part 1: What is the problem? What is the topic of this paper?

Part 2: How is the problem solved (methodology)?

Part 3: What are the specific results? How well is the problem solved?

Part 4: So what? How useful is this to science or to the reader?

---

[b] *Ibid.*

A four-part abstract should be the norm. However, many have only three parts: the fourth one (the impact) is missing. Why?

1. Was the maximum number of words allowed by the journal reached too quickly because a long rambling start justified the importance of the problem, thereby forcing the author to skip or reduce a part?
2. Did the author (mistakenly) consider that the results speak for themselves?
3. Could it be that the author was not able to assess the impact of the scientific contribution, a result of the myopia caused by the atomisation of research tasks among many researchers?

Whatever the reason, having less than four parts reduces the informative value of the abstract and, therefore, its value to the reader. Since the reader decides whether to read the rest of your article or not based on the abstract, its incompleteness reduces your chances to be read and cited.

Before studying the abstract in greater detail, it is necessary to identify the author's contribution from the title of the paper. Where does the bar | separating the contribution from the context go?

"Nonlinear finite element simulation to elucidate the efficacy | of slit arteriotomy for end-to-side arterial anastomosis in microsurgery"

In the abstract, the parts that cover the contribution should be more developed. In this abstract, they correspond to parts 2 through 4. Did you notice a discrepancy between title and abstract in this sample paper? There is one. If one evaluates the contribution by the number of words for each part, it seems that part 3, the elucidation of the efficacy, is the contribution (112 words). Part 2, the nonlinear finite element analysis, plays an incidental role (only 10 words). The title could have been the following:

> *Elucidating the efficacy | of slit arteriotomy for end-to-side arterial anastomosis in microsurgery with a nonlinear finite element simulation*

However, after examining the structure of the paper (headings and subheadings), it appears that the contribution is indeed the nonlinear finite element simulation. The title is therefore correct. One concludes that the abstract is aimed at surgeons who care little about the technical details of the contribution, but more about the surgical method and its efficacy. Had the paper been targeted towards computer scientists, the methodology part would have been longer and the results part shorter. The readers of the *Journal of Biomechanics* in which this paper was published come from very diverse horizons. In both cases, however, the parts relative to the contribution contain the largest number of words (140–150 out of 211 words).

 **Read your abstract and locate its various parts. Does your abstract have its four essential parts? Are the parts with the largest number of words those corresponding to the contribution? Are you still using adjectives in the results section, or have you given enough precision?**

Coherence Between Abstract and Title

A rapid calculation will determine whether an abstract is coherent with its title. In this calculation, articles, (*a, an, the*, etc.) and prepositions (*of, on*, etc.) are not taken into account. In the example above, 5 (41%) of the 12 significant title words are both in the title and in the first sentence of the abstract. This percentage is good. Why?

It really is a matter of common sense. Your title creates an expectation: the reader, having read the title, expects to know more about it as soon as possible. Can you imagine an abstract disconnected from the message of its corresponding title? It is unimaginable. The coherence between title and abstract is achieved through the repetition of words. Percentages outside the 30%–80% range should be examined more closely.

0%. There could be a problem. The first sentence deals with generalities loosely related to the topic of the paper. EXCEPTIONALLY, one sentence of background may be written to set the problem in its context. This is part zero of your abstract. Totally optional, it should be the exception, not the rule. In any case, it should at least contain one word from the title.

20%. The first sentence contains one or two title words. It sets the background to the problem, or briefly explains one or two unusual title keywords. This is fine, as long as sentences 2 and 3 mention most of the other title words. Otherwise, the background is too long and, as a result, the abstract lacks conciseness.

90%–100%. Idyllic percentage? Not necessarily. The first sentence is often a straightforward repetition of the title with just a verb added. Why repeat? The first sentence should expand, not just repeat, the title. However, if it contains many more words than the title, then 100% may be acceptable.

To summarise, the first sentence of your abstract should contain at least one third of the words in your title (these words are frequently found in the second part of your title, i.e. its context). Your title merely whets the appetite of your readers; they expect to know more about your title in your abstract. You should satisfy their expectation and rapidly provide more precise details.

First count the total number of significant words in your title (do not include small words, such as *on, the,* or *a* in your count). Let's call that number *T*.

Then, identify in your first sentence the significant words that are also in the title. Underline these words IN THE TITLE. Modified forms (a noun changed to a verb or vice versa) are acceptable but synonyms are not. For example, *simulation* would be considered the same as *simulated,* but *abrasion* would not be the same as *corrosion.*

Count the number of words underlined in your title. Let's call that number *U*. Finally calculate the percentage $100 \times U/T$.

What is your percentage? Between 30 and 80%, you are doing fine. Outside of this range, investigate.

A second calculation will help you identify the strength of the cohesion between abstract and title. Are ALL title words also in the abstract? They should be. Think about it. You give high visibility to a word by giving it "title" status — the highest status in a paper. Why would title words be missing in the abstract? It may be for the following reasons:

1. You used the synonym of a title word to avoid repetition. Why? By doing so, you miss out on a great opportunity to reinforce the message communicated in the title. Repeating a title word in the abstract will also increase the relevance score calculated by search engines for that keyword. As a result, your title will be brought up towards the top of the list of titles retrieved. Using an alternative keyword is acceptable only if two keywords are interchangeably used in your field. The alternative keyword would then increase the probability that your title is found by search engines.

2. The title word is not important. Remove it from the title to increase conciseness.
3. The title word missing in your abstract is really important. Find a place for it in your abstract.
4. It may, also be that your abstract contains a keyword that should be in the title, but is not. In that case, rewrite your title to incorporate that keyword.

You have already calculated T the number of significant words in your title. Read your abstract and see if any of the important title words are missing. If some are, ask yourself why. It may be that your title claims are too broad, your title is not concise enough, you are using synonyms that dilute the strength of your keywords and confuse the reader, etc. Decide which reason applies, and modify the title or abstract if necessary. If you are yet to write your first paper, use the sample abstract (arteriotomy).

You now have four techniques to gauge the quality of your abstract.

1. Abstracts have four parts. The part that represents your contribution should be the most developed.
2. Abstracts repeat their title words in full. (A possible exception to this recommendation is when you use alternative keywords because a particular concept is expressed by two equally probable keywords and you want your paper to be found/retrieved. You then use one keyword in the title, and the other equally probable keyword in the abstract.)
3. Abstracts expand the title in the first two or three sentences because the reader expects it.

4. Abstracts need to set the problem, but do not need to justify why it is important (the introduction does that). They need, however, to justify the significance of the results (*a posteriori* impact).

The Tense of Verbs in an Abstract

Abstract written in the present tense only?

Never say never! I used to think that abstracts were only written in the past tense because it refers to work that is completed. It looks as though I was mistaken. More journals now accept the use of the present tense for an abstract. Before you select the tense for your abstract, read the journal recommendations to the author. If the journal does not forbid the use of the present tense, you may want to consider it for the reasons given below. Once you have chosen a tense, keep it throughout your abstract.

There are added advantages to choosing the present tense for the abstract. The present tense is vibrant, lively, engaging, leading, contemporary, and fresh. The past tense is passé, déjà vu, gone, stale, unexciting, and lagging. It feels like reading old news. The researcher has finished a Herculean task and describes it without excitement, as a thing of the past. Furthermore, the past tense can create ambiguity. For example, the phrase *was studied* creates doubt: did the writer publish this before? Remember that the reader still needs to be convinced to read your WHOLE paper. So far, you have managed to bring him past the title barrier and into your abstract. Congratulations! But if the reader stops there, you might never get to make an impact on science. You do want the reader to plunge inside your paper. The abstract is going to act as a diving board. Your accomplishments are better communicated through the use of the dynamic present tense than the dull past tense. Here is an additional final reason for using the present tense in your abstract if the journal accepts it. The part of a paper often read after the abstract is ... the conclusions. Alas, the conclusions are also written in the past tense. To the reader, reading the conclusions feels like reading the abstract all over again (boring).

Purpose and Qualities of Abstracts

Purpose of the abstract for the reader

1. It makes the title clear.
2. It provides details on the writer's scientific contribution.
3. It helps the reader decide whether the article is worth reading or not.
4. It helps the reader rapidly gather competitive intelligence.
5. It helps the reader assess the level of difficulty of the article.

The abstract is NOT to be used for the following:

1. To mention the work of other researchers (it is the role of the introduction), except when your paper is an extension of a (one) previous paper, yours or that of another author.
2. To justify why the problem you have chosen is important (it is also the role of the introduction). Your abstract should concentrate on the importance of the results, not that of the problem.

Purpose of the abstract for the writer

1. It allows the paper to be found more easily, because it has more keywords than the title.
2. It states the writer's contribution in more precise detail than the title (adjectives in the title are frequent, but they should be rare in the abstract).

You could also write two abstracts: one put together before starting your paper to capture the gist of the contribution, and the other written after your paper is complete to capture the heart and soul of the paper. The two may differ, for they serve different purposes: one guides, the other summarises.

Qualities of an abstract

An abstract is **COMPLETE.** It has four parts (what, how, results, impact).

An abstract is **TIED TO TITLE.** All title words are found in the abstract.

An abstract is **CONCISE.** It is not longer than necessary, as a courtesy to the reader. Justification of research is best done through significant results.

An abstract is **STAND-ALONE.** It lives by itself in its own world: databases of abstracts, journal abstracts. It needs nothing.

An abstract is **REPRESENTATIVE** of the contribution of the paper. It sets expectations for the reader.

An abstract is preferably **PRESENT.** Real. News.

Not all abstracts have four parts, sometimes with good reason. A review paper that covers the state of the art in a particular domain has only one or two parts. Short papers (letters, reports) have one or two lines. "Extended" abstracts are written prior to a conference, in some cases before the research is even completed; as a result, their parts 3 and 4 are shallow or missing. But, apart from these special cases, all abstracts should have four parts.

What do you think of your abstract? Does it have enough of the qualities mentioned here? Is the contribution you mention in your abstract consistent with that claimed by the title? A quality abstract makes a good first impression. Spend some time reviewing it.

12

Headings/Subheadings: The Skeleton of Your Paper

The skeleton gives a frame to the body. With it, the reinforced body takes shape; without it, the human would be a jellyfish. The skeleton of a paper is its structure. *The skeleton supports the various parts of the body according to their functional needs.* Composed of headings and subheadings set in a logical order, the structure reinforces the scientific contribution. *The skeleton is standard, but it allows for variations in shape and size.* Headings are generally the same from one article to the next (introduction, discussion, conclusions), but subheadings differ. *The most sophisticated parts of the skeleton are also the most detailed (backbone, metacarpus, metatarsus).* The most detailed part of a structure contains the largest amount of contributive details.

The scientific paper: 300 years of history

In an article published in *The Scientist* entitled "What's right about scientific writing", authors Alan Gross and Joseph Harmon[a] defend the structure of the scientific paper against those who claim that it does not represent

(Continued)

---

[a] Gross A and Harmon J, "What's right about scientific writing", *The Scientist* 13:20, 1999.

> *(Continued)*
>
> the way science "happens". The structure, refined over more than 300 years, has enabled readers to evaluate the trustworthiness and importance of the presented facts and conclusions. The authors praise the standard narrative. In addition, they observe that today, as a result of the increased role played by visuals, it is necessary to go beyond the interpretation of linear text.

Three Principles for a Good Structure*

A structure that plays its role follows these principles:

1. The contribution guides its shape.
2. Title words are repeated in its headings and subheadings.
3. It tells a story clearly and completely in its broad lines.

Studying the structure of your paper will allow you to identify important problems. Your paper may be too complex, too detailed, too premature, or too shallow.

Let us review the structure of the paper on slit arteriotomy. You should now be familiar with the title and abstract of this paper. In the structure that follows, words in italic type are common to both title and structure.

Nonlinear *finite element simulation* to elucidate the efficacy of *slit* arteriotomy for end-to-side *arterial* anastomosis in microsurgery

1. Introduction
2. Mechanical factors underlying *slit* opening

---

*These principles do not apply to review papers. Principle 1 is also not applicable for journals that impose a set detailed structure for their methodology section (in chemistry and biology for example) to enable independent verification of experiment.

3. Methodology for computer *simulation*

    3.1. Reference configuration for the *finite element* model

    3.2. Geometry details and boundary conditions of *the finite element* model in the reference configuration

    3.3. Hyperelastic material for the *arteries*

    3.4. *Simulation* procedure for the operation

4. Results and discussion

5. Conclusions

References[b]

Recall that the title is composed of two parts: the front part reflects its contribution; and the back part, its context.

"Nonlinear finite element simulation to elucidate the efficacy" [**Contribution**] of slit arteriotomy for end-to-side arterial anastomosis in microsurgery [**Context**] "

Principle 1: the contribution guides the shape of a structure

In the example above, three headings are standard: "*Introduction*", "*Results and discussion*", and "*Conclusions*". Standard headings are disconnected from titles, since they contain no title word. They are simply marks that indicate the location and function of a part. In contrast, headings 2 and 3 are more meaningful: they contain nearly half the title words.

Headings 1 and 2 cover the background. They have no subheadings. Headings 4 and 5 present the results and conclusions. They also have no subheadings.

---

[b] Reprinted from Gu H, Chua A, Tan BK, and Hung KC, "Nonlinear finite element simulation to elucidate the efficacy of slit arteriotomy for end-to-side arterial anastomosis in microsurgery", *J Biomech* **39**:435–443, 2006 (with permission from Elsevier).

Heading 3 dominates this structure. With four subheadings, it provides much detail on the contribution. The subheadings organise the details in a logical order. All of this is to be expected, is it not? A structure should be the most detailed where the author has the most to write about, namely the scientific contribution of the paper. The structure has to expand to match the level of detail by offering more subheadings to help organise these details in a logical order, for the benefit of the reader and for the sake of clarity (☞1).

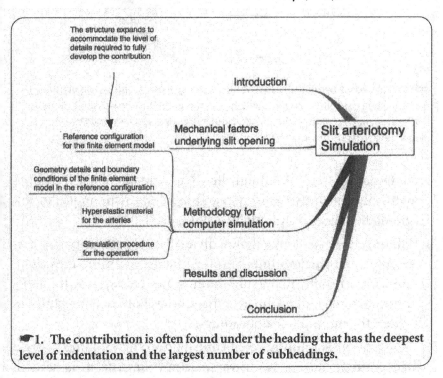

☞1. The contribution is often found under the heading that has the deepest level of indentation and the largest number of subheadings.

This first principle has a corollary: *when excessively detailed parts do not contain much contribution, the structure has a problem.*

1. A secondary part may be overly detailed. Simplify or put details in appendices or footnotes.

2. The knowledge level of the reader is underestimated. Remove details and provide references to seminal papers and books (☞2).

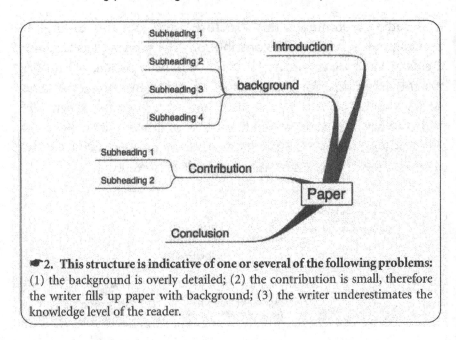

☛2. This structure is indicative of one or several of the following problems: (1) the background is overly detailed; (2) the contribution is small, therefore the writer fills up paper with background; (3) the writer underestimates the knowledge level of the reader.

3. Subheadings are "sliced and diced" too small. When a section with only one or two short paragraphs has its own subheading, it should be merged with other sections.

4. The top-level structure is not divided into enough parts. For example, the background section is merged with the introduction. As a result, many subheadings are necessary within the introduction. Add headings at the top-level of your structure to reduce the number of subheadings.

5. The paper has a multifaceted contribution that requires a large background and an extensive structure. Rewrite it as several smaller papers (☛3).

Principle 2: title words are repeated in the headings and subheadings of a structure

Is it reasonable to imagine a structure disconnected from its title? Since the role of a structure is to help the reader navigate inside your paper and identify where your contribution is located, a structure should have its headings and subheadings connected to the title.

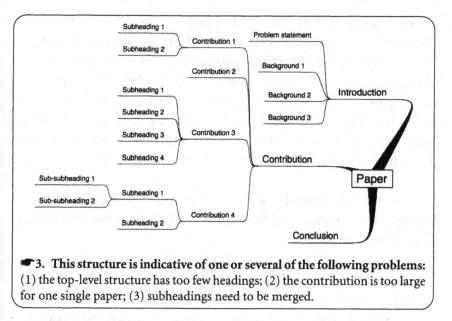

☛3. **This structure is indicative of one or several of the following problems:** (1) the top-level structure has too few headings; (2) the contribution is too large for one single paper; (3) subheadings need to be merged.

Let us apply this second principle on our sample structure and consider headings 2 and 3.

2. *Mechanical factors underlying **slit** opening*

Heading 2 contains "*slit*", a title word found in the second half of the title describing the context. Therefore, heading 2 is unlikely to be about the contribution of the paper. It extends the introduction and provides additional background to the reader, namely the surgery steps and the mechanically induced stresses and deformations observed during the surgery, because these will be modelled and analysed under heading 3.

3. *Methodology for computer **simulation***

    3.1. *Reference configuration for the **finite element** model*
    3.2. *Geometry details and boundary conditions of the **finite element** model in the reference configuration*
    3.3. *Hyperelastic material for the arteries*
    3.4. ***Simulation** procedure for the operation*

Heading 3 and its four subheadings contain "*simulation*" and "*finite element*", two words located in the front part of the title (contribution part). They confirm that this heading covers the contribution of the paper. The author could have added "*nonlinear*" to strengthen the coherence between title and structure. The specificity of the words in the heading and subheadings immediately conveys to the non-computer expert that this section of the paper is very technical. This structure is clear to computer programmers, but less so to surgeons.

This second principle has a corollary: **when headings and subheadings are disconnected from the title of a paper, the structure or the title may be wrong.**

1. The title of the article may not be the right one. The structure reflects the contribution better than the title.

> **The wrong title**
>
> I remember examining a paper where the word "trajectory" was present in three of the five headings, yet it was totally absent in the title. One gets suspicious!

2. The structure is too cryptic. Its headings and subheadings are too generic, brief, or tangential. They do not give enough information on the contents. Revise the structure and reconnect it to the title.

3. Synonyms replace keywords. Having lost homogeneity and coherence, the article is less clear. Return to the original keywords.

Principle 3: a structure tells a story that is clear and complete in its broad lines

According to this third principle, someone unfamiliar with the domain of computer simulations should be able to see the logic of the story after reading the title, the abstract, and the successive headings and subheadings.

Is this story clear?

1. *Introduction*
2. *Mechanical factors underlying slit opening*
3. *Methodology for computer simulation*

    3.1. *Reference configuration for the finite element model*
    3.2. *Geometry details and boundary conditions of the finite element model in the reference configuration*
    3.3. *Hyperelastic material for the arteries*
    3.4. *Simulation procedure for the operation*

4. *Results and discussion*
5. *Conclusions*

Heading 2 paints the landscape. The reader enters the operating theatre, and observes the surgeon cut and stitch the arteries. They open under the sharp blade of his scalpel, and deform under the pressure of his fingers and the pull of the stitches. One can imagine, once the surgery is completed, the blood flowing through the artery, opening the slit wider.

Heading 3 provides details on the contribution: a simulation. 3.1 defines the initial state of the simulated objects. 3.2 gives details on the model parameters (arteries, slit) and defines their limits. 3.3 describes how the arteries, key objects in the simulation, will be modelled. 3.4 makes the simulation steps correspond to the steps of the actual surgery.

The story is coherent with what the title announces, but it is incomplete. There is no link between the model and the result (elucidation). This could easily be achieved by replacing the standard heading "*Results and discussion*" with a more informative heading such as "*Elucidation of the efficacy of slit arteriotomy*", thus establishing a clear direct connection between the model and its results.

The third principle has a corollary: ***when headings or subheadings read in sequence tell a nonsensical story, the structure has a problem.***

1. The paper could be premature: its structure has not yet reached clarity. More work is needed until the structure falls into place. The story is not ready yet.

2. The story is nonsensical because it is not the story of the title, but another story. Change the title or rewrite the paper. You have the wrong face for the right body, or vice versa.

3. The headings and subheadings are too cryptic. Write more informative headings and subheadings.

Syntactic Rules for Headings

Traditionally, and to help the reader rebuild a story from its structure, headings at the same indentation level or subheadings under the same heading adopt a parallel syntax. In the model structure, headings 2 and 3 are noun phrases. Within heading 3, all subheadings are also noun phrases.

In the following structure, however, the syntax is not parallel.

1. Introduction
2. Interference mechanism
3. Design rules
***4. Proposing** a solution*
4.1. Three-layer prediction algorithm
    4.1.1. Algorithm classification
    4.1.2. Layer prediction comparison
5. Proposed recognition
6. Simulation studies
7. Discussion
8. Conclusions

This is not a good structure for many reasons. Focusing solely on the lack of consistency, one cannot miss the "one parent and only one child" problem: heading 4 has only one subheading 4.1 (no 4.2). The syntax also lacks consistency at the same heading level: headings 1, 2, 3, 5, 6, 7, and 8 are all single-noun phrases; but heading 4 starts with a present participle "*Proposing*", thus breaking the consistency (or parallelism in syntax).

Purpose and Qualities of Structures

Purpose of the structure for the reader

> 1. It makes navigation easy by providing direct access to parts of the paper.
>
> 2. It helps the reader locate the section of the paper related to the author's contribution.
>
> 3. It allows the reader to quickly grasp the main story of the paper by making a logical story out of the succession of headings and subheadings.
>
> 4. It sets reading time expectations through the length and detail level of each section.

Purpose of the structure for the writer

> 1. It reinforces the contribution by repeating key points or achievements in the headings or subheadings.
>
> 2. It helps the writer divide the paper into informative sections that support the contribution (Some writers use structure as a framework for writing. They create the structure, and then write. This method has value. It gives focus to the paper. If the story flows well at the structural level, then it will probably flow well at the detailed level as well. You may still change the

structure as you write, but it will mostly be to refine the headings or to create more subheadings, not to totally restructure the flow of your paper.).

Qualities of a structure

A structure is **INFORMATIVE.** No empty signposts are found outside of the expected standard headings. The contribution is clearly identified in the nonstandard headings.

A structure is **TIED TO TITLE AND ABSTRACT.** Keywords from the title and abstract are found in the structure. They support the contribution.

A structure is **LOGICAL.** Between headings, and within each heading, the reader sees the logic of the order chosen by the writer.

A structure is **CONSISTENT** at the syntax level. Each parent heading has more than one child subheading. Syntax is parallel.

A structure is **CONCISE.** Neither overly detailed nor too condensed, the structure helps the reader discover the essential.

Not all papers have an explicit structure. When the paper is short (e.g. an IEEE letter), the structure is implicit. The "*Introduction*" heading is absent, but the first paragraph of the letter introduces and the last one concludes. Also, not all papers have verbless headings. In some journals, each heading is a full sentence.

Here is a very simple and productive method to ascertain the quality of your structure. "Flatten" your structure on a blank piece of paper. By this, I mean write the title at the top of the page, and then write ALL headings and subheadings in the order they appear in your paper. Once done, underline the words that are common in the structure and in the title. Do you detect any discrepancy here? Are words from the title missing in the structure? Should words from the structure be part of the title? Is your structure very disconnected from your title?

Once you have examined how well the structure matches your title, have someone else read your flattened structure and explain to you what he or she thinks your paper contains. The less this person knows about your work, the better. Ask this person if the logic is visible in the succession of headings or subheadings. If the person is largely puzzled, you are not quite ready to publish yet. Rework your structure and your paper. When the story is clear, give a quick syntactic check. Is the syntax of your headings parallel? Are subheadings orphans?

When the volunteer reviewer asks questions, do not start explaining! Remember that the reader will not be there for you to explain once your paper is published. Just take note of the observations, and correct the structure or title accordingly.

13

Introduction:
The Hands of Your Paper

Extended hands welcome. They invite to enter and guide someone unfamiliar with a new place. The introduction of a paper plays a similar role. It provides guidance, greets, and introduces a topic not familiar to the reader. *Hands point to something worthy of attention, and invite the eyes to follow.* The introduction also points to the related works of other scientists and to your contribution.

What Is Wrong with a Short Boilerplate Introduction?

For many, the introduction is a necessary evil, something more difficult to write than the methodology or results section. Therefore, to ease the burden, the scientist usually keeps it short and builds a three-part introduction: a concise obligatory introductory paragraph to describe and justify the problem, a brief related works section to place the contribution in context with densely packed references, and a final paragraph to formally introduce the main headings that follow. Alas, the brevity and lack of detail are only appreciated by the few experts in the field who are already familiar with the introduction material. The *many* readers with a significant knowledge gap will not be satisfied.

How many is "many"?

Just travel back in time to the last conference you attended. Visualise yourself browsing through the conference schedule and its many concomitant presentations competing for your attention. Do you recall which sessions you chose? Naturally, you selected the presentations immediately relevant to you, those that were right in your field of research. But, you also attended others, even though you did not quite have all the necessary background to fully understand them. They looked intriguing and potentially helpful. You were not alone to venture into the interesting unknown. Based on the survey I regularly conduct, on average, 30% to 50% of the presentations that scientists attend are slightly outside their field.

It is safe to say that *many* scientists (a reasonable 40%) will require an introduction to your paper. Could reviewers be among them? They could be. Therefore, write an introduction that will bridge their knowledge gap, otherwise they will not be able to evaluate your paper correctly. Remember that they have veto power over the selection of your paper for publication.

What do readers expect from an introduction? Three readers give their opinion here. The underlined words are worthy of your attention.

Xiaoyan

"*I want to know clearly what the objectives and the motivations are. I expect the author to justify his or her research.*" The use of the word "*clearly*" implies that, even after reading the title and abstract, some readers are still unclear about the author's objectives.

Mary

"In the introduction, I expect to find the context, the background, what others are doing in this field, things like that. I also want to know what really is new in the paper. If the introduction is well written, I usually read the rest of the paper." The title and abstract only provide a hint to what is new in the paper. A clearer understanding of what is new will have to wait until the reader has finished reading the related works section, so as to identify how the contribution differs from the work of others.

Kumar

"I don't usually read introductions. Most of what's in there is repeated verbatim elsewhere in the paper anyway. They are a waste of time. They always say the same thing: the problem is important, everybody else but the author is doing it wrong, and they usually end with a boring table of contents. So, I skip them." People who have read too many bad introductions can easily identify them.

The comments of Kumar and Mary reveal that it is not just **what** you put in your introduction that matters, but also **how** you write it. This chapter and the next will review the content and style of an introduction.

The Introduction Answers Key Reader Questions

Imagine scientists reading the first lines of your introduction. They have identified your title as containing something of interest. They may have ordered or downloaded your paper. It is now on their desk or on their computer screen. They have just read your abstract and understand your contribution, but not in detail. Writers often believe that after reading the title and abstract, readers should have a clear picture of their contribution. However, this is not the case. A key ingredient is lacking in the dry, disembodied abstract: the context or background. Therefore, the **first** duty of the writer is to **briefly establish the context**.

Here is an example taken from life sciences.

Name entity recognition (NER), an information extraction task, automatically identifies named entities and classifies them into predefined classes. NER has been successfully applied to newswires [references]. Today, researchers are adapting NER systems to extract biomedical named entities — protein, gene, or virus [more references] — for applications such as automatic build of biomedical databases. Their success is limited.

After reading this paragraph, the reader expects the writer to explain why success is limited, and to bring an answer to the main question "What adaptations to NER will enable biomedical named entities to be extracted more successfully?"

What is the main question of your paper? It is the question to which your contribution or the title of your paper is the answer. If you cannot phrase your contribution in question form, then you are not ready to write your paper because you do not yet have a clear idea of your contribution. To help you determine the main question, practise on the following familiar titles:

"Nonlinear finite element simulation to elucidate the efficacy of slit arteriotomy for end-to-side arterial anastomosis in micro-surgery"[a]

---

[a] Reprinted from Gu H, Chua A, Tan BK, and Hung KC, "Nonlinear finite element simulation to elucidate the efficacy of slit arteriotomy for end-to-side arterial anastomosis in microsurgery", *J Biomech* **39**: 435–443, 2006 (with permission from Elsevier).

Main question:

Why does slit arteriotomy work so well?

"*Energy-efficient data gathering in large wireless sensor networks*"[b]

Main question:

How can a sensor node be chosen to forward data in a large network so that total energy consumption for the data gathering is minimum?

Read your title and abstract. Write the main question they answer. Is this question clearly stated in your introduction? If there is more than one question, you may have a paper with multiple contributions, and possibly a paper that could be divided into multiple papers. Alternatively, you may not yet clearly understand your contribution.

Now that you know the main question, include it in your introduction as soon as you can. It helps reviewers and readers understand the problem in a clear, attention-grabbing, and succinct way. It even helps you to remain focused. Naturally, the main question triggers many others.

---

[b] Lu KZ, Huang LS, Wan YY, and Xu HL, "Energy-efficient data gathering in large wireless sensor networks", *Second International Conference on Embedded Software and Systems (ICESS'05)*, Xi'an, China, pp. 327–337, 2005.

> ### The questionable cake
>
> One afternoon, Vladimir Toldoff received a call from his wife Ruslana as he was finishing an experiment in the lab. "I am coming with one cake, two plates, and assorted cutlery", she announced. He answered, "What? Wait! First, what is the occasion? And why now? Can't it wait until tonight? And by the way, what cake is it, and why do you want to cut it in the lab? You know that crumbs are not welcomed here."
>
> The rapid fire of questions did not faze Ruslana. She knew her Vladimir. A full-fledged scientist. She paused and rephrased his questions succinctly. "All right, let me see. You would like to know why a cake, why eat it now, why its mouth-watering taste should make you shout 'Darling, come right away', and why I should slice it in the lab instead of at home. Am I right?" Vladimir, quite impressed with her matter-of-fact answer, started to laugh. "That's right", he responded. Ruslana then uttered three words that had him shout for joy: "My Medovik cake."

Similar questions are asked by the reader of a scientific article as shown hereunder (ignore the initials and domain terms, and concentrate instead on the story thread).

1. **Why now?** In this case, because previous studies produced conflicting results.

   > "_We were curious to see whether <u>we</u> could resolve the discrepancy between these gene profiling studies by using <u>our</u> current understanding of the gene differences between GCB and ABC DLBCL._"[c]

---

[c] Wright G, Tan B, Rosenwald A, Hurt E, Wiestner A, and Staudt LM, "A gene expression-based method to diagnose clinically distinct subgroups of diffuse large B cell lymphoma", *Proc Natl Acad Sci U S A* **100**(17): 9991–9996, 2003. © 2003 National Academy of Sciences, USA.

2. **Why this?** In this case, because it was challenging.

   "As was pointed out (3), it is a challenging task to compare the results of these profiling studies because they used different microarray platforms that were only partially overlapping in gene composition. Notably, the Affymetrix arrays lacked many of the genes on the lymphochip microarrays...."[d]

3. **Why this way?** In this case, because it worked with different platforms.

   "For this reason, we developed a classification method that focuses on those genes that discriminate the GCB and ABC DLBCL subgroups with highest significance."[e]

4. **Why should the reader care?** In this case, because it predicted survival.

   "Our method does not merely assign a tumor to a DLBCL subgroup but also estimates the probability that the tumor belongs to the subgroup. We demonstrate that this method is capable of classifying a tumor irrespective of which experimental platform is used to measure gene expression. The GCB and ABC DLBCL subgroups defined by using this predictor have significantly different survival rates after chemotherapy."[f]

Readers rely on you to answer these fundamental questions.

The reviewer has another set of questions. Even though they overlap with the scientific reader's questions, they differ in some ways.

1. Is the problem good and is solving it useful?

2. Is the solution new, clear, and effective compared to others?

---

[d] *Ibid.*
[e] *Ibid.*
[f] *Ibid.*

3. Is the solution the best one for this problem?

4. How does this paper help the readers of the journal?

Therefore, you should have both reader and reviewer in mind when you write your introduction. It is up to you to convince them that the problem is real, and that your solution is original and useful.

The Introduction Sets the Foundations of Your Credibility

A solution that is claimed to be universal and better than any other is not very credible. I remember reading an online article on presentation skills[8] that claimed that if only one side of an issue is presented, then believeability is in the low 10%; but if both sides are presented (of course, the negative side is only presented after the good side has had ample opportunity to be discussed), then believeability is in the high 50%. The title of this particular slide was "fairness". In science, it would have been "intellectual honesty".

Intellectual honesty is demonstrated in many ways. One of them cannot be ignored by the author: a clear and honest description of the of problem's scope and the solution's application domain. Readers need to know the scope of your work because they **want** to benefit from it; therefore, they need to evaluate how well your solution would work on their problems. If the scope of your solution covers their area of need, then they will be satisfied. If it does not, at least they will know why, and they may even be encouraged to extend your work to solve their problems. Either way, your work will be helpful.

> **The drug info sheet**
>
> To be really scared, do not go and see a horror movie. Instead, go into your medicine cabinet, and read the
> (*Continued*)

---

[8] Broker J, "Persuasive presentations: tips for presenters", http://www.uccs.edu/rjbroker/bio401/handouts/persuasive %20 presentations.doc

(Continued)

piece of paper folded in eight sandwiched between the two strips of aluminium holding the precious pills that may cure your headache. Take the time to read the microscopic text to build up some really unhealthy anxiety. The warnings are so overwhelming that if the pills do not cure you, they might just as effectively lead you straight to the emergency room.

If the pharmaceutical companies disclose these limitations, it is to avoid lawsuits and to help doctors prescribe the right medicine. Not stating limitations in your scientific paper is unlikely to kill anyone, but it may damage your reputation — a reputation based on honesty as well as results. You decide: are your results good in spite of restrictive limitations or because of them? The reader needs to know.

Scope

In essence, the scope of your contribution is carved by your method, hypothesis, and data. Establishing a frame around the problem and the solution enables you to claim, with some authority and assurance, that your solution is "good" inside that frame. Some writers leave the framing until later in the paper, usually in the methodology section, because they are afraid of discouraging the reader. However, I believe that a reader informed early on the scope is better than a reader disappointed by the late disclosure of restrictive assumptions and limitations that unexpectedly restrict the applicability and value of your work. Therefore, establish the scope early in your paper.

Warning: Not all assumptions affect the scope. Specific assumptions are better mentioned just-in-time in the paragraphs in which they apply (often in the methodology section). Justify their use or give a measure of their impact on your results, as in the following three examples: (1) *Using the same assumption as in [7], we assume that . . .;* (2) *Without loss of generality, it is also assumed that . . .;* and (3) *Because we assume that the event is slow varying, it is reasonable to update the information on event allocation after all other steps.*

The choice of method is also best justified in the introduction to strengthen the credibility of your work, as the following examples illustrate.

Our dithering algorithm does not make any assumption on the resolution of pictures, nor does it make any assumption on the colour depth of the pixels.

Our method does not need to consider a kernel function, nor does it need to map from a lower dimension space to a higher dimension space.

Definition

Another way of framing is by defining. In the following example, the authors define what an *"effective"* solution is. They do not let readers decide the meaning of this adjective.

An effective signature scheme should have the following desirable features:

1. *Security: the ability to prevent attacked images from passing verification;*
2. *Robustness: the ability to tolerate incidental distortions introduced from the predefined acceptable manipulations such as lossy compression;*
3. *Integrity: the ability to integrate authentication data with host image rather than as separate data; and*
4. *Transparency: the embedded authentication data are invisible under normal viewing conditions.*

When you define, you frame, i.e. you restrict the meaning of the words to your definition. Demonstrating that a solution is "good" because it fulfils predefined criteria is easier than demonstrating that a solution is "good" when the evaluation criteria are left up to the reader.

To conclude, a good story is a story that one can believe. Precisely because they want to benefit from your work, scientists critically question what they read. Can they believe everything that is written? The slightest doubt in their mind will cast doubt on the rest of the paper. Further down in the paper, in the discussion section, they will accept suggested explanations from the author, but only if they have been convinced right from the start.

So far, only two ways to establish credibility have been presented: scope and definition. In the next chapter, two other ways will be considered: citations and precision.

The Introduction Is Active and Personal

The analysis of Mary and Kumar's views revealed that the way an introduction is written is just as important as what is in it. The introduction is the place to write about your findings and your reasoning in story form. Because this story is about you, make it lively, engaging, and personal. Use pronouns such as *we* or *our*. Do not follow those who claim that it is improper to mention yourself. Poor Vladimir Toldoff listened to them and found out that they are not always right.

> **The story of Vladimir Toldoff**
>
> "Vladimir!"
> The finger of Popov, his supervisor, is pointing at a word in the third paragraph of Vladimir's revised
> *(Continued)*

(Continued)

introduction. "You cannot use 'we' in a scientific paper. You are a scientist, Vlad, not Tolstoï. A scientist's work speaks for itself. A scientist disappears behind his work. You don't matter, Vlad. 'The data suggest' . . . you cannot write 'our data'. It's THE data, Vlad. Data do not belong to you. They belong to science! They speak for themselves, objectively. You, on the other hand, will only mess things up, and introduce bias and subjectivity. No Vlad, I'm telling you: stick to the scientific traditions of your forefathers. Turn the sentences around so that you, the scientist, become invisible. Write everything in the passive voice. Am I clear?"

"Crystal," Vladimir responds, "But I was only taking the reviewer's comments into account." With that, he hands out the letter he recently received from the editor of the journal. His supervisor grabs the letter impatiently.

"What kind of nonsense is this?" (reading the letter aloud) . . . *Your related work section is not clear. You write, "The data suggest". Which data? Is it the data of [3], or is it your data? If you want me to assess your contribution fairly, you should make clear what YOUR work is and what the work of others is. Therefore, if it is your data, then write, "our data suggest". Also, if I may make a suggestion, I feel that your introduction is somewhat impersonal and hard to read. You could improve it by using more active verbs. That would make reading easier . . . .*

"Ah, Vladimir! No doubt this comes from a junior reviewer. What is happening to science!"

Often times, a paper is a collective effort. Therefore, refer to yourself using *we. I* is suitable for professors or Nobel Prize winners who write alone.

Let us look again at an earlier example. Notice the very personal tone of the paragraph, as well as the use of the active voice.

> "**We** were **curious** to see whether **we** could resolve the discrepancy between these gene profiling studies by using **our** current understanding of the gene differences between GCB and ABC DLBCL."[h]

The story of the passive lover

Imagine yourself at the doorstep of your loved one. You are clutching, somewhat nervously, a beautiful bouquet of fragrant roses behind your back. You ring the doorbell. As your loved one opens the door and gives you a beaming smile, you hand out the bouquet of flowers and utter these immortal words:

"You are loved by me."

What do you think happens next?
(a) You eat the flowers; or
(b) You ring the doorbell again and say the same thing using the active voice.

The passive voice is quite acceptable in the rest of your paper, where who does what does not really matter. In the introduction, however, the passive voice has a dampening effect. The introduction is the story of the "what's" and the "why's"; it is a story, not a report. This is the one place in the whole paper where you, as a writer, can relax and write in a way very close to the way you would write to a friend, your friend the reader, to whom you offer your contribution in the hope that it will be useful.

---

[h] Wright G, Tan B, Rosenwald A, Hurt E, Wiestner A, and Staudt LM, "A gene expression-based method to diagnose clinically distinct subgroups of diffuse large B cell lymphoma", *Proc Natl Acad Sci U S A* **100**(17): 9991–9996, 2003. © 2003 National Academy of Sciences, USA.

The Introduction Is Engaging and Motivating

The introduction engages and motivates readers to read the rest of your paper. After reading it, they must be "fired up", wanting to know more. If everything goes well, readers will appreciate you as a writer, not just as a scientist. Do you remember Kumar's views on the introduction?

> *"I don't usually read introductions. Most of what's in there is repeated verbatim elsewhere in the paper anyway. They are a waste of time. <u>They always say the same thing</u>: the problem is important, everybody else but the author is doing it wrong, and they usually end up with a <u>boring</u> table of contents. <u>So, I skip them.</u>"*

Kumar thinks that introductions are often boring and repetitive. Why repetitive? Are they rewritten several times for several journals, losing a little of their flavour each time? Are they copied from the introduction of other researchers working in the same field? Why boring? Is it because they are written after the work is finished, after the fun and the excitement have gone? This is why writing the introduction of your paper early in your project is good. You still have the excitement of the journey that lies ahead to energise your writing: the tantalising hypothesis, the supportive preliminary data, and the fruitful methods.

A slow introduction start, particularly the "vacuous" and the "considerable" starts, will delay and bore the reader.

The vacuous false start

> *In the age of genomes, large-scale data are produced by numerous scientific groups all over the world.*

> *Significant progress in the chemical sciences in general, and crystallography in particular, is often highly dependent on extracting meaningful knowledge from a considerable amount*

of experimental data. Such experimental measurements are made using a wide range of instruments.

Because of the long-term trend towards smaller and smaller consumer goods, the need for the manufacture of microcomponents is growing.

Was there anything in these examples you did not already know? Catch and ruthlessly destroy these cold starts, these hollow statements where the writer warms up with a few brain push-ups before actually getting down to the matter at hand. You will be more concise.

Here is another false start, even though it tries to conjure up excitement through the sheer size of the problem (not the solution).

The considerable false start

There has been a surge, in recent times, towards the increasing use of . . .

There has been considerable interest in recent years in this technology, and, as trends indicate, it is expected to show continuing growth over the next decade . . .

In this type of false start, the author considers the heat of a research field sufficient to warm up the reader. The words used are symptomatic: *exponential, considerable, surge, growing, increasing.* The readers, however, used to these excessive claims, remain ice-cold and their eyes skip the verbiage. An important class of readers, the reviewers, will immediately suspect a "me-too" paper: the writer is obviously running behind the pack. Many people may consider the problem important, but that does not make your contribution an important one.

It is best to start with what readers expect: an explanation of the problem mentioned in the abstract, and a description of its context. This guarantees conciseness.

Fireworks usually end with a bang. Introductions should end likewise, and the bang is the foreseable impact of your contribution. Alas, too often, the ending of an introduction is flat. Here are typical lacklustre endings.

The dead end

> *The rest of this paper is organised as follows: section 2 discusses related works. Section 3 presents the technology, and shows how our approach is conducted using our scheme. Section 4 presents the results of our experiments, and shows how the efficiency and accuracy of our approach compares with others. Finally, we offer our conclusions and discuss limitations.*

> *The rest of this paper is organised as follows. Section 2 describes some related works, in particular similar work that has been done. Following that, the proposed approaches are discussed in section 3, with the implementation details being discussed in section 4. Section 5 evaluates the performance, and compares the proposed approaches to a baseline model. Finally, we draw conclusions and outline future works in section 6.*

These table-of-content endings have no place in an introduction, except in large documents where readers cannot just flip a few pages and discover the whole structure. Michael Alley[i] advocates "mapping the document in the introduction". He gives the example of a journal article where the author successfully manages to present in story form an overview of the methodology, thereby answering the "why this way?" question. The story reveals the problem and the method used to solve it. It remains silent on the results, but indicates the impact of being able to solve the problem.

---

[i] Alley M, *The Craft of Scientific Writing*, Springer, New York, 1997.

Is it best to keep the reader in suspense or should your introduction end on a restatement of your contribution and its foreseeable impact? To answer this question, consider what the reader has read before your introduction: your title and abstract. One could argue that since the reader already knows your contribution, it is not necessary to repeat it. Withholding it in the introduction does not create a suspense that will draw the reader to your paper. The reader already has your paper in hand! You have already successfully taken the reader through the filters of the title and abstract.

Therefore, why tell the end of the story in the introduction instead of keeping that for the conclusions? Simply because the abstract and title do not place your contribution in the context of the work of others. Presenting your contribution in the context of other people's work does not need to wait till the conclusions; it can happen right there and then in the introduction of your paper. Repeating the contribution in different settings strengthens and clarifies it. A word of warning, however: do not get caught copying and pasting sentences from various parts of your paper into the introduction. The readers who remember having read the same words elsewhere will not appreciate your hurry.

See how the abstract and the introduction differ in the following paper (again, ignore the acronyms and focus on the similarities and differences).

Abstract

"The GCB and ABC DLBCL subgroups identified in this data set had **significantly different 5-yr survival rates after multiagent chemotherapy (62% vs. 26%; P ≤ 0.0051), in accord with analyses of other DLBCL cohorts.** These results demonstrate the ability of **this gene expression-based predictor** to classify DLBCLs into biologically and clinically distinct subgroups irrespective of the method used to measure gene expression."[j]

Introduction

"**We** demonstrate that this method is capable of classifying a tumor irrespective of which experimental platform is used to measure gene expression. The GCB and ABC DLBCL subgroups defined by using **this predictor** have **significantly different survival rates after chemotherapy.**[k]

The abstract is more precise than the introduction when it comes to the key numerical results. But, the factual abstract does not tell a personal story: "*These results demonstrate*" is passive, whereas "*We demonstrate*" is active.

[j] Wright G, Tan B, Rosenwald A, Hurt E, Wiestner A, and Staudt LM, "A gene expression-based method to diagnose clinically distinct subgroups of diffuse large B cell lymphoma", *Proc Natl Acad Sci U S A* **100**(17): 9991–9996, 2003. © 2003 National Academy of Sciences, USA.
[k] *Ibid.*

Read the first paragraph of your introduction. Is it vacuous or considerable? If it is, delete it. Is the last paragraph redundant with the structure? If it is, delete it. Did you use the pronoun "we"? Did you answer all the "why's"? Identify each "why" in your text. Read chapter 5 again on bridging the knowledge gap. Did you use your scientific logbook to build interest in the story of your introduction? If you did not, why not? Did you cut and paste parts of your abstract into your introduction or vice versa? If you did, rewrite. To identify whether or not you adequately scoped your problem and solution, simply underline the sentences that deal with scope, assumptions, and limitations. Are there enough of them? Are they at their proper place? Finally, did you mix the introduction with technical background? If you did, create a separate heading after the introduction for your background. The introduction captures the mind; the background fills it. These two functions are best kept separate.

14

Introduction Part II: Popular Traps

The introduction helps the reader to understand the context from which your research originated. Other scientists also work in your field. Do you borrow or adapt their work to reach your objective, or do you follow a completely different research path? The reader wants to know. Positioning your work on the research landscape is a perilous exercise because it is tempting to justify your choices by criticising the work of other scientists.

Four traps are laid in the path of writers: the trap of the story plot, the trap of plagiarism (using someone else's words without proper quotes and acknowledgement), the trap of imprecision, and the trap of judgmental adjectives.

The Trap of the Story Plot

The introduction tells the personal story of your research. All stories have a story plot to make them interesting and clear. Here is a frequent story plot found in introductions.

> **A story**
>
> I'm so excited about telling you this great story. My father [1] □ is on the front lawn cleaning the lawn mower. My sister [2] △ is in the back kitchen making a cake. My mum [3] ★ has gone shopping, and I ○ am playing my electric guitar in my bedroom.
>
> Do you like my story? No??!!! It's a great story. What's that you're saying? My story has no plot??? Of course there is a plot! See, it describes my family's activities, starting with my father. We all have something in common: family ties, living under one roof, etc. Here we are again. □ △ ★ ○

If this story left you cold, the analogous story found in scientific papers will also leave the reader cold. In short, the story says the following: in this domain, this particular researcher did this; that research lab did that; in Finland, this other researcher is doing something else; and I am doing this particular thing. The problem with this type of story is that the relationship between their work and your work is not stated. In symbolic graphical form, the story plot would look like ☞1. The pieces are juxtaposed, not linked.

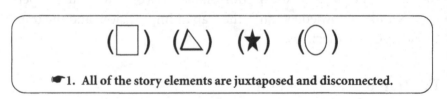

☞1. **All of the story elements are juxtaposed and disconnected.**

Contrast the first story with this one:

> **A better story**
>
> I'm so excited. I'm going to tell you a great story. My father [1] □ is on the front lawn cleaning the lawn
> *(Continued)*

(Continued)

mower. And do you know what that means? Trouble! He hates it. He wants everyone to help bring this or bring that in order to feel less miserable. Whenever that happens, we all run away, not because we refuse to help him, but because he wants us to stand there and watch idly while he works. So, my sister [2] △ is taking refuge in the back kitchen and is plunging her hands in flour to slowly make a cake. My mum [3] ★ has suddenly discovered that she is missing something or other, and has rushed out to shop, saying she'll be gone for an hour or so. As for me, ○ I am in my bedroom playing the electric guitar with my amplifier at maximum volume.

A thread that links all the parts together is necessary to make an interesting story plot, as in ☛2. The difference is striking, is it not? The story has a direction and its parts are all connected.

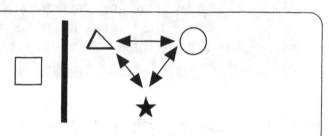

☛**2. Three story elements share a common bond.** This bond isolates the square element.

Here is a second story based on a story plot frequently found in scientific papers.

A terrible story

I'm so excited. I'm going to tell you my second best story. A red Ferrari [1] ○ would take me to Vladimir

(Continued)

(Continued)

Toldoff's house in 5 hours. It is fast. **However**, it is very expensive [2,7,12]. A red bicycle [3] ✧ is much less expensive and is quite convenient for short trips. So, if Vladimir Toldoff came to live near my house, it would be quite cost-effective [4]. **However**, a bicycle that doesn't have a mudguard requires a bicycle clip [5] so as not to dirty trousers. Since red athletic shoes [6] ✦ do not require a bicycle clip, they are a better solution than a bicycle to travel short distances [8]. **However**, their colour is easily degraded [9] by adverse weather conditions, particularly in the muddy rainy season. **On the other hand**, brownish open plastic sandals [10] ◯ do not have any of the previous problems: they are cheap, convenient, require no bicycle clip, and do not show mud stains. Furthermore, they are easy to clean, and are fast to put on. **However**, contrary to the Ferrari, they reflect poorly on the status [11] of their owners. Therefore, I am working on a framework to integrate self-awareness into the means of transportation, and will validate it through the popular Sims 2 simulation package.

Yes, I have exaggerated (only a little), but you get the point. The *however* plot, after taking readers through four sharp *however* turns, completely loses and confuses them. The seemingly logical connection between the elements is tenuous, as in ☞3. On the way to the last proposal (the writer's contribution), a long list of disconnected advantages and disadvantages is given; by the time readers get to the end of the list, they innocently (and wrongly) assume that the final solution will provide all the advantages and none of the disadvantages of the previous solutions. Unfortunately,

☞**3. Story elements loosely connected two by two.** Four shapes: a sun, a star, a cross, an ellipse. The first element is compared with the second, the second with the third, and so on. At the end, the final element is connected back to the original element, thus completing the loop. Yet, the sun is never compared with the cross, and the star is never compared with the ellipse. For the chain of comparisons to be meaningful, the comparison criteria must be identical for all elements, and all elements must be compared.

the comparison criteria continuously vary, and therefore nothing is really comparable.

Both plots are frequently found because they are convenient from a writer's perspective:

1. They allow a list of loosely related references to be easily assembled.
2. The shallow analysis of related works is fast because it does not require extensive reading of other people's work (abstracts or titles are enough in most cases).

Are there better plots? Assuredly. But, giving examples would fill too many pages. That is why I have presented them in schematic form in ☞4.

I have found that a plot that works well in movies is also useful in scientific writing. The author shows you how the story ends before it even starts. When readers have the full picture, they are better able to situate your work in it. They understand how and with whose help you will achieve it. In addition, they know your limitations and expect that, in the future, you will deal with them. Graphically, this story plot is represented in ☞5.

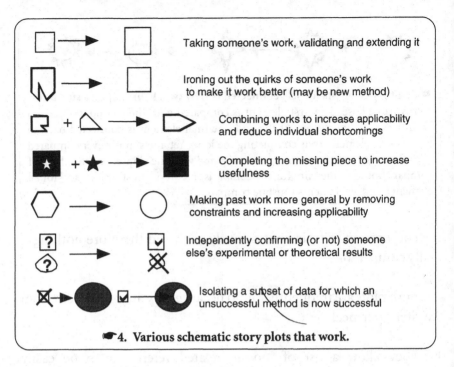

Taking someone's work, validating and extending it

Ironing out the quirks of someone's work
to make it work better (may be new method)

Combining works to increase applicability
and reduce individual shortcomings

Completing the missing piece to increase
usefulness

Making past work more general by removing
constraints and increasing applicability

Independently confirming (or not) someone
else's experimental or theoretical results

Isolating a subset of data for which an
unsuccessful method is now successful

☞4. **Various schematic story plots that work.**

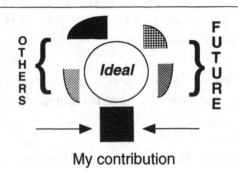

My contribution

☞5. **The ideal solution story plot.** First, the ideal system or solution is depicted (circle). Then, the story tells how this ideal picture comes together: what the author contributes (square); what others have already contributed (left brace); and what still remains an open field of research, but with its parts identified (right brace). Everything is clear, everything fits nicely, and the reader is more easily convinced of the worth of your contribution.

Identify your story plot. Does it look like a series of juxtaposed elements without any connection or like a "however" meander? Is your story easy to follow? Does it flow logically: from past to recent, from general to specific, from specific to general, from primitive to sophisticated, from static to dynamic, from problem to solution, or from one element of a sequence to the next in line?

The Trap of Plagiarism

Plagiarism exists when someone else's words are found in your paper without proper quotes **and** references. Plagiarism is a taboo subject in many research centres. Senior researchers, whose names often appear as the third or fourth author in a paper, do not need to be told. Their reputation is at stake. They know only too well the hefty price one pays when caught. They have heard the tale of the faculty dean high up in the research ladder who had to resign because someone found out that he had plagiarised in a paper he wrote 20 years earlier, while he was still a junior researcher.

> **Vladimir Toldoff told off again**
>
> "Vladimir!"
>
> The finger of Popov, his supervisor, points to a sentence in the third paragraph of Vlad's introduction in the paper published 3 months earlier in a good journal.
>
> "Yes, anything wrong?"
>
> "The English in this paragraph about Leontiev's algorithm is too good. These are not your sentences."
>
> "Um, let me see. Ah, yes, it is rather good, isn't it! I must have been in great writing shape that day. I remember
>
> *(Continued)*

(Continued)

noticing how well I had worded that paragraph when I cut and pasted it into my paper from my reading notes."

"Would it be too much to ask you to bring your reading notes?"

"You have access to them already. I left you the files after the review meeting last month."

"Oh yes. That's right. Let's have a look on my computer ..... Here are your notes on Leontiev's work. Vlad, this looks like a 'cut and paste' segment to me, a cut and paste segment without proper quotes."

"Are you sure?" Vladimir asks.

"It's easy to see; let me retrieve Leontiev's paper from the electronic library. Just a minute. Here it is. Let me copy a sentence from your paragraph and do a string search on Leontiev's paper and ... well, well, well! What do we have here?! An exact copy of the original!"

"Oh NO!" Vladimir turns red. But, he recovers quickly and smiles widely. "It's fine! Look! I put a reference to Leontiev's work right at the end of the paragraph. A reference is the same as a quote, isn't it? After all, Leontiev should be happy. I am increasing his citation count. He will not come and bother me by claiming that these words are his, not mine."

Popov remains silent. He retrieves from the top of his in-tray basket what looks like an official letter and reads it out loud.

"Dear Sir,
One of my students has brought to my attention that a certain Vladimir Toldoff who works in your research cen-
tre has not had the courtesy to quote me in his recent
(Continued)

> *(Continued)*
> *paper, but instead has claimed my words to be his (see paragraph 3 of his introduction). I am disappointed that a prestigious institute like yours does not carefully check its papers prior to publication. I expect to receive from your institute and from Mr Toldoff a letter of apology, with a copy forwarded to the editor of the journal.*
> *I hope this is the last time such misconduct will occur.*
> *Signed, Professor Leontiev."*

When plagiarism occurs, it is often due to a less than perfect methodology to collect and annotate the background material. Keeping relevant documentation about the information source when capturing information electronically is simply good practice.

Plagiarism is very subtle. One may think that by changing a word here and there, one gets rid of plagiarism. However, this is not the case. Literature has a term for this bad practice: "patchwork plagiarism". One may also think that one does not plagiarise if one changes every word except the keywords in successive sentences (thus keeping the meaning of these sentences identical). Again, this is not the case. What is protected is not just the words, but also the succession of ideas in successive sentences. Indeed, if I translated a passage in French, all of the words would be different, but I would still be plagiarising. The ideas expressed in successive sentences would be exactly the same. In short, the only way to avoid plagiarising is to completely rewrite without looking at the original document, or to restructure ideas and add value by reordering them according to a different point of view: your point of view.

Even subtler is plagiarism of oneself. You might think that it is unnecessary to quote a sentence from one of your earlier publications. But, is this really the case? Did you write the paper alone, or were there coauthors? How would they feel if their work was not recognised?

Furthermore, you are often asked to assign your copyright to the journal, in which case the reproduction rights of your article no longer belong to you. Copying large chunks of your past publications (including visuals) would constitute a breach of copyright, unless it is authorised.

All publishers have someone dealing with permissions. I had to send countless e-mails and faxes to receive the permission to quote the examples used in this book. Some publishers responded promptly, while others took more than a month. Some publishers allow you, in their contract, to republish your article in whole or in parts (for example, on your website), but there are always restrictions. To retain more rights, some writers pay a publication fee (about US$1500 per article in 2007) to publish in open access journals. The added advantage is that open access articles appear to be cited more often (see http://opcit.eprints.org/oacitation-biblio.html).

The lure and anonymity of the web may be so tempting that sentences, even visuals, get copied here and there. However, free or open access does not imply free right of use for everyone. Sometimes, data, text corpora, photos, and video frames may be available online to allow researchers to benchmark their algorithms; but unless permission to reproduce is granted by the owners of the benchmark, copying these in their original form is not legal, even in situations where you copy only part of the data or image.

Quoting is good practice. Adding quotes shows that you have read the papers you refer to or compare. When you give credit where credit is due, you have everything to gain and nothing to lose. Science is the fruit of collective work. Quoting scientists who have been published, particularly if they are well respected, adds credibility to your own work. It makes your work more authoritative. If you do not share their views, quoting what you object to cannot be disputed. You do not interpret; you cite.

Observe how Professor Feibelman quotes others.

"In apparent support of the half-dissociated overlayer, Pirug, Ritke, and Bonzel's x-ray photoemission spectroscopy (XPS) study of $H_2O/Ru(0001)$ 'revealed a state at 531.3 eV binding energy which is close to [that] of adsorbed hydroxyl groups' (28)."[a]

Note how skilfully he quotes from another report, while at the same time he hints (using the word *"apparent"*) that the support is not there at all. Indeed, the next sentence (not shown here) starts with *"However"* and confirms the lack of support.

Plagiarism has becoming such a problem that most journals are now using plagiarism-detection software (*Turnitin, Copyscape, Ithenticate, Crosscheck*) to detect scientists who plagiarise. Research institutes that value their reputation often require researchers to check their paper against plagiarism prior to submission. It is only a matter of time before such checks are conducted retroactively. Woe to the researchers found plagiarising, even twenty years ago.

The Trap of Imprecision

This is another trap you could easily fall into if you were in a hurry. Your paper may mention 20 or more references. Under the pressure of a conference- or manager-imposed deadline, you may be tempted to prepare the related works section from abstracts, not from the full text of papers that you did not have time to read. There may be other reasons to justify this behaviour: the library may not have the paper and would need to order it, the paper may not be available online, or someone else in the group (now on vacation) may have borrowed it.

---

[a] Reprinted excerpt with permission from Feibelman PJ, "Partial dissociation of water on Ru(0001)", *Science* **295**:99–102, 2002. © 2002 AAAS.

☛6. **Words that are indicators of a lack of precision in scientific writing.**

Typically	*A number of*	*Several*	*Many*	*Most*
Generally	*The majority of*	*Less*	*Others*	*A few*
Commonly	*Substantial*	*Various*	*More*	*The main*
Can /May	*Probably*	*frequent*	*Often*	*...*

The consequences of incomplete reading are far from negligible, in particular if reading goes no further than the abstract (or worse, if it stops at the title). Abstracts do not contain *all* of the results, they do not mention assumptions or limitations, and they do not justify the methods used. As a result, your sentences will resemble this: "*Many people have been working in this domain [1,2,3,4,5,6,7,8,9,10], and others have recently improved what their predecessors did [11,12,13,14,15,16,17].*" Reviewers will see through the smokescreen.

Abstract skimming, or dotting your paper with references of articles you have not read, will hurt you in many ways.

1. Errors will creep into your paper.
2. Because they find your domain knowledge too superficial, reviewers will be tempted to lower the value of your contribution.
3. Your research will not be clearly positioned on the research landscape.
4. Your story will lack detail and, therefore, interest.
5. The reader will doubt your expertise because your words lack assurance. Readers are usually quick to detect authors who write with authority from the level of details and precision in their paper. Remember the devastating effect of doubt.

If any of the words from the table in ☛6 are found in your introduction, then you may have fallen into the trap of imprecision.

Read your introduction, and circle the words you find in the list in ☞6. Do you need them? How authoritative are you? Can you delete them, or replace them with more specific words or numbers to increase precision?

The Trap of Judgmental Adjectives

Some adjectives and adverbs are dangerous when used in the related works section of your paper. The danger comes from their use in judgmental comparisons. Adjectives such as *poor, good, fast, faster, not reliable, primitive, naïve,* or *limited* can do a lot of damage. They make your work look good at the expense of others who came before you. These very people may one day read what you have written about them, and will understandably be upset.

Does this mean that all adjectives are bad? No, they are just dangerous. Every adjective is a claim; and in science, claims have to be substantiated. How would you explain and justify the adjective *poor* if it refers to the performance of a system?

What adjectives (if any) are to be used? Adjectives you justify later with data or visuals, adjectives that compliment (with reason) authors of related works, and adjectives that reflect undisputed public knowledge. Let adjectives be based on facts, or on quotes from other authors stating their own limitations or assumptions.

Here are four ways to avoid direct judgment:

1. State that your work agrees (disagrees) with another paper's conclusions, or state that your results are coherent with (different from) those found in another paper.
2. Use facts and numbers (quantitative instead of qualitative comparisons).

3. Define your uniqueness, your difference (nothing is comparable to what you do).

4. Quote another paper that independently supports your views.

In his book *Reglas y Consejos sobre Investigación Científica: Los tonicós de la voluntad,* Santiago Ramón y Cajal[b] recommends indulgence because methodology is the source of many errors. He never doubts that the author has talent, commenting that if the author had access to the same equipment he used, he or she would have arrived at the same conclusion. In any case, the author was published and his efforts contributed to the advancement of science, whether they were crowned with success or not.

Cajal's book was published at the end of the 19th century. His words have aged, but not his kindness. Let us respect other people's work.

Read your introduction, and underline the adjectives you find a little too judgmental or gratuitous. Replace them with facts or citations.

Purpose and Qualities of Introductions

Purpose of the introduction for the reader

1. It brings the reader up to speed and reduces the initial knowledge gap.

2. It poses the problem, the proposed solution, and the scope in clear terms.

3. It answers the questions raised by the title and the abstract.

---

[b] Ramón y Cajal S, *Reglas y Consejos solve Investigación Científica: Los tonicos de la voluntad,* López-Ocón L (ed.), Gadir Editorial, Madrid, 2005.

Purpose of the introduction for the writer

1. It gives the writer a chance to loosen the tie, unbutton the collar, and write in a personal way to the reader.
2. It sets readers' expectations for the yet unread part of the paper, and enhances (or not) their motivation to find out more in the rest of the paper.
3. It showcases the writer's expertise in communication skills, scientific skills, and social skills.
4. It enables the writer to strengthen the contribution.

Qualities of an introduction

An introduction is **MINDFUL**. The author makes a real effort to assess and bridge the knowledge gap.

An introduction is **STORY-LIKE**. It has a plot that answers all the "why" questions of the reader one by one. It uses the active voice and includes the writer ("we"). Verbs are conjugated using various tenses: present, past, future.

An introduction is **AUTHORITATIVE**. References are accurate and numerous, comparisons are factual (not judgmental), related works are closely related, and imprecise words are absent.

An introduction is **COMPLETE**. All "why's" have their "because". The key references are mentioned.

An introduction is **CONCISE**. No considerable or vacuous beginnings, no table-of-content paragraphs, no excessive details in answering the "why's", no historical panegyric.

15

Visuals: The Voice of Your Paper

A voice attracts attention; it announces, it warns. It is a substitute to writing: one can read a book or listen to a recorded version of it. Likewise, photos and graphics shout their messages, sometimes without any words. They are worth a thousand words. *The voice gets out of the body. It is not necessary to see the body to hear its voice.* Visuals inform readers independently, even before they start reading the first paragraph. *A voice has its own language, a universal and wordless language, like the one used by the child who babbles, laughs, and cries.* Visuals have their own language: the universal language of graphic arts. They tell a story directly and quickly with a minimum of text. *Voice intonation reinforces the message expressed by the body.* Visuals also reinforce the main message of the text, and are in synergy with it.

Just observe the title of this chapter for a few seconds, and then bring your eyes back to the text. Headings and subheadings shout, don't they? They are so authoritative in their bold font suit. Framed by a white space, nothing crowds them in their spacious surroundings. The reader understands them at a glance.

Tables and diagrams speak just as much as photos. Guided by a grid of vertical and horizontal lines, bold font, and arrows, the reader captures a large volume of information in little time and easily extracts trends and relationships between the visual elements. The visual story is told in a few words.

Visuals excel in visualising the results of your scientific work. Admire their talents in ☛1.

☛1. Functions of Visuals.	
To compare and contrast	To represent complexity
To give precision and detail	To provide context
To summarise	To reveal sequence
To classify	To reveal patterns
To establish relationships	...

Readers prefer a visual to text for several reasons:

1. Readers explore the visual actively. They keep the initiative. They probe with silent questions and look for answers in the visual, its title, or its caption.
2. Readers enjoy the speed, the freedom, and the challenge of exploration. Guided by the intrinsic logic within the visual, their eyes leap from one place of interest to another; whereas in linear text, their eyes walk.
3. Between the text part (title and caption) and the graphic part of the visual, there is redundancy. Because of it, readers understand more easily.

Visuals have a loud and convincing voice, but only if you can make them speak. Their language is based on a special grammar that describes the correct use of fonts, blocking, kerning, framing, white space, line and colour, etc. Grammatically correct visuals have impact and are highly readable. Unfortunately, the visual language is well understood mostly by graphic designers. They can make a visual shout, whereas most of us can only make it whisper or croak. This chapter is not about graphic design, but rather about the correct use of visuals in a scientific paper. It is also about principles that will help you design visuals that have an impact from a scientific perspective, even if the lines are a little thin, the white space is not quite well distributed, or the kerning is an abomination. You may not get an Oscar in a design competition, but you will have visuals that do more than whisper or croak.

Seven Principles for Good Visuals

After reading hundreds of papers, I have detected consistent patterns among bad visuals. A bad visual breaks one or more of the following principles.

1. A visual does not ask more questions than it can answer.
2. A visual is custom-designed to support the contribution of only one paper.
3. A visual keeps its complexity in step with readers' understanding.
4. A visual is designed based on its contribution, not on its ease of creation.
5. A visual has its elements arranged to make its purpose immediately apparent.
6. A visual is concise if its clarity declines when a new element is added or removed.
7. Besides the caption, a visual requires no external text support to be understood.

Principle 1: a visual does not ask more questions than it can answer

Before you set your eyes on ☞2, let me predict what will happen: I am confident that, within a few seconds, your brain will predictably direct your eyes to key parts of the visual; probing, evaluating, and asking silent questions. The first questions will be "what-am-I-looking-at" questions.

It is now time to put your eyes to the test on ☞2, but do not look for the title and caption of this figure because I have removed them to help you focus on the graphic.

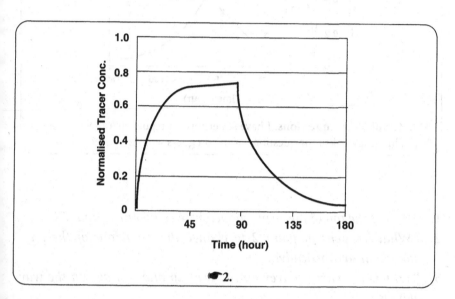

☞2.

"*What is that curve? What are on the axes?*" Time (hour) is on the x-axis; whatever the curve is, a 100-hour process is slow! Concentration (I think "Conc." means that) is on the y-axis. So, this graph shows the evolution of the concentration of a tracer (probably fluorescent) over time.

As soon as these questions are answered, others surface after the brain recognises familiar patterns/templates and singles out points of interest. Your eyes return to the curve. This curve looks like a quadratic curve, but the differences need to be explained. Look at ☞3.

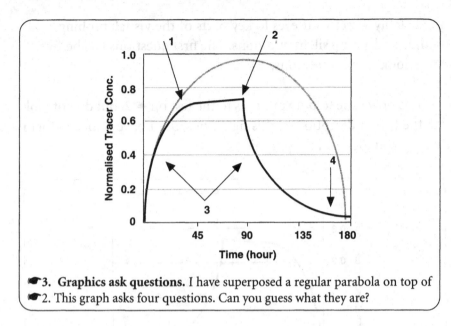

☞3. **Graphics ask questions.** I have superposed a regular parabola on top of ☞2. This graph asks four questions. Can you guess what they are?

1. *"Why is the top of the curve clipped between points 1 and 2?"*
2. *"What happens at point 2 to change the behaviour of the phenomenon so drastically?"*
3. *"Why is the curve convex on the way up and concave on the way down?"*
4. *"Why is the curve asymptotic for high values of time?"*

The readers will look for these answers in the caption. If they are left unanswered, then readers will be frustrated.

The screenshot is widely used to illustrate scientific papers because a mouse click is all it takes to capture it. However, this type of visual frequently raises more questions than it answers, as seen in the screenshot in ☞4. It includes all the artefacts of the software application: menu items, windows, icons, tool palettes, and other distracting elements that raise questions.

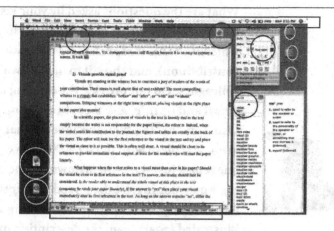

☞4. **Visual gallery of errors exhibit: the cluttered screen 'dump'.** If my objective is to show the content of the centre window, then what are the circled and framed elements doing in this visual?

Hence, the first principle: a visual does not ask more questions than it can answer. To discover what these questions are, ask a reader and be prepared for surprises.

1. Readers come up with unexpected questions caused by their lack of initial knowledge, or by the unfamiliar acronyms or abbreviations used.
2. More unexpected questions arise from bad visual design where important details are missing, or from distracting details that defocus the reader away from the main point of your visual.

Once the questions are known, you have two choices: you could answer them directly in the caption, the title, within the visual itself, or indirectly by providing additional background in the body of the paper; or you could trim the figure to focus on the essential contribution, without raising distracting questions. The only choice you do **not** have is to ignore the questions.

What are the questions asked by your visuals? Choose the key visual in your paper (the one which is the most representative of your contribution), and show it to one or two colleagues. Do not provide the caption; only the visual and its title. Ask them to question your visual and make a hypothesis as to what you want to achieve with it. DO NOT answer their questions; just write them down. At the end of their questioning, let them read the caption. Ask if the caption contains information not illustrated (apart from the description of the visual's context or the visual's interpretation). If it does, remove the information in excess or illustrate it in the visual. Show your modified visual to new colleagues and verify that the caption now answers ALL questions.

Principle 2: a visual is custom-designed to support the contribution of only one paper

When you created a particular visual (long before your paper was written), it was a work of art. Using Photoshop, you or the graphic artist had spent a lot of time to make it look perfect. It had attracted much attention at an internal technical presentation or at a previous conference. Part of it does illustrate a key point in your paper, but that part cannot be easily extracted from your masterpiece without serious reworking of the original. You are tempted to recycle the whole drawing/diagram. As a result, the visual includes much information that does not "belong" to your paper: names, curves, numbers, or acronyms foreign to the reader. All raise questions. Hence, the second principle: a visual is custom-designed to support the contribution of only one paper.

Redrawing is a small price to pay for a custom-designed visual in view of the benefits, not just to the readers, but also to you: (1) you do not have to ask for the permission to reuse the original, if the reproduction rights belong to the journal where it first appeared; and (2) your contribution is easier to identify because it is not drowned in a sea of irrelevant details.

Principle 3: a visual keeps its complexity in step with readers' understanding

Visuals are star witnesses standing in the witness box to convince a jury of readers of the worth of your contribution. Their placement in your paper is as critical as the timing lawyers choose to bring in their key witness. More importantly, their convincing power is far beyond that of text exhibits. The most compelling witness is a visual that compares "before" and "after", or "with" and "without". Comparisons require visuals to be side by side.

To place a complex visual in a paper, one has to take into account the level of understanding of the reader. More complex visuals can be placed closer to the end of the paper, when the reader understands more. Simpler visuals can be placed anywhere.

Your visual could be placed at an inappropriate location in two situations: when others (e.g. the staff in charge of the layout for the journal) decide on its placement, or when your visual is referred to more than once in the body of your paper.

When you send your paper to a journal, its figures and tables are usually at the back of the paper after the references (unless you submit the paper as a PDF file). The editor will look for the first reference to the visual in your text, and will try to place it as close to its reference as possible. This is often well done. However, if you want to make sure that your visual is properly placed, limit its width to one or several columns of the journal in which you intend to publish your paper and avoid using small font sizes that cannot be made smaller without reducing readability.

What happens when you refer to a visual more than once in a paper? Should the visual be close to its first reference in the text or its last? An obvious answer would be the former. However, assuming readers read your paper linearly from introduction to conclusions, if the visual is placed close to its first reference, readers may find it too complex because they have yet to acquire the knowledge that will make the visual totally understandable.

So, before answering this question, consider a more fundamental question: why is it necessary to refer to the visual more than once? Is it because you are making multiple points in one large or complex visual? If this is so, divide the one complex visual into several parts [(a), (b), (c)] to reduce its complexity. Next, make sure that readers have enough information to understand *everything* in Figure 1(a) when reaching Figure 1(a) in the text, and do likewise for parts (b) and (c). However, if after dividing the visual into parts you realise there is no value in having visuals 1(a), 1(b), and 1(c) next to one another, say for comparative reasons, then divide the visual into separate visuals and have them appear in a just-in-time fashion next to their respective references.

If, as recommended in principle 7 (hereafter), your complex visual is self-contained, you can refer to it many times in the paper without any problem.

Principle 4: a visual is designed based on its contribution, not on its ease of creation

Information with visual impact requires creativity, graphic skill, and time. Because most of these are in short supply, software and hardware producers provide creativity, skill, and time-saving tools: statistical packages that crank out tables, graphs, and cheesy charts in a few mouse clicks; digital cameras that, in one click, capture poorly lit photos of experimental setups replete with noodle wires (I suppose the more awful they look, the more authentic they are); and screen capture programs that effortlessly lasso and shrink your workstation screen to make it fit in your paper. The ease of creation of visuals contributes to their abundance — mouse-produced becomes mass-produced. This abundance, however, may have unexpected side-effects.

When I ask researchers to read a paper and indicate what the key graph/figure/table representative of the contribution is, their choice often differs from the author's choice. Why? It may be due to the author's inability to make his or her contribution visually clear, but the cause may also lie elsewhere. The more visuals you have, the more likely your contribution will be diluted across them, and the more difficult it will be for the reader to grasp your whole contribution succinctly. This unpleasant side-effect hides another one: when the time comes to cut down the number of visuals (that time will inevitably come), if your contribution is dispersed between them, removing some will lessen total understanding. You will need to consolidate (redo, merge, redesign) the visuals and link them again to your text. This will take time. Therefore, be selective and design your visuals based on how much they showcase your contribution, not on how easy they are to create.

If I showed you a photo of a keyboard with the caption "keyboard on which this book was typed", as in ☞5, would it contribute greatly to the usefulness of this book? Such photos are frequently found in scientific papers, but they are not useful. They only prove that the writer conducted an experiment using real equipment. To ensure that each visual is critical to your paper, ask yourself whether it replaces much text or strongly supports your contribution. Conciseness applies to both text AND visuals. Verbose visuals can defeat a good paper.

To summarise, if verbose circumstantial evidence (which fails to convince a jury) dominates your paper at the expense of succinct

☞5. Gallery of errors exhibit: a QWERTY keyboard, but who wants to know! This photo speaks volumes, doesn't it? It tells you that I use a Macintosh PowerBook with a titanium casing, that I do not use a French keyboard even though I am French, that my right shift key is broken in two, and finally that I am not much of a photographer! What does this have to do with the book itself? Nothing much.

but detailed convincing evidence, the following will occur: (1) your contribution will be diluted; (2) the reader will be at a loss to identify the key visual; (3) you will lose time redoing visuals when asked to shorten your paper; and (4) your article will not be concise.

How many visuals do you have in your paper? Could you identify the one that encapsulates your major contribution? Could other people? Are you verbose or concise when it comes to visuals? What do your readers think?

Principle 5: a visual has its elements arranged to make its purpose immediately apparent

Complexity is present when the brain cannot easily identify or connect related elements. To create complexity is easy: simply bury the key information in the midst of data.

The visual salvo is a popular classic in the gallery of errors leading to complex visuals (☞6). The visual is impressive, but the reader is not

6. Visual gallery of errors exhibit: the visual salvo. A large number of visuals are set side by side; each one is marginally different from the one that precedes it, so much so that the eye can barely see the difference between them. In this case, the visuals are tables, but it could also be graphs or images.

$\approx$	PostAt15(liquid)		PostAt30(liquid)		PostAt30(solid)		PostAt30(solid)	
	$\beta = 1$	$\beta = 20$	$\beta = 1$	$\beta = 20$	$\beta = 1$	$\beta = 20$	$\beta = 1$	$\beta = 20$
B4	0.5323+17.4%	0.5323+18.9%	0.4225+19%	0.4254+20.0%	0.2157+9.6%	0.2185+11.1%	0.1493+9.3%	0.1501+8.4%
B6	0.5323+17.4%	0.5373+18.9%	0.4202+18.0%	0.4254+20.1%	0.2156+9.5%	0.2171+10.8%	0.1493+9.0%	0.1500+8.5%
B8	0.5324+17.5%	0.5373+18.9%	0.4189+17.7%	0.4255+20.0%	0.2156+9.5%	0.2182+10.9%	0.1492+9.1%	0.1496+8.5%
BJI	0.4720	0.4720	0.3706	0.3706	0.2997	0.2997	0.2380	0.2380

Table 1 Statistics on β − 1 or 20, LG = 3

$\approx$	PostAt15(liquid)		PostAt30(liquid)		PostAt30(solid)		PostAt30(solid)	
	$\beta = 1$	$\beta = 20$	$\beta = 1$	$\beta = 20$	$\beta = 1$	$\beta = 20$	$\beta = 1$	$\beta = 20$
B4	0.5456+23.4%	0.5423+23.9%	0.4324+26.0%	0.4341+27.0%	0.2192+12.6%	0.2232+14.9%	0.1544+13.3%	0.1554+14.8%
B6	0.5440+23.4%	0.5473+23.9%	0.4332+25.0%	0.4341+27.1%	0.2193+12.5%	0.2235+14.1%	0.1544+13.3%	0.1550+14.6%
B8	0.5424+23.5%	0.5473+24.0%	0.4299+24.7%	0.4341+27.0%	0.2126+12.6%	0.2231+14.7%	0.1543+13.0%	0.1556+14.0%
BJI	0.4720	0.4720	0.3706	0.3706	0.2997	0.2997	0.2380	0.2380

Table 1 Statistics on β−1 or 20, LG = 4

$\approx$	PostAt15(liquid)		PostAt30(liquid)		PostAt30(solid)		PostAt30(solid)	
	$\beta = 1$	$\beta = 20$	$\beta = 1$	$\beta = 20$	$\beta = 1$	$\beta = 20$	$\beta = 1$	$\beta = 20$
B4	0.5400+22.1%	0.5323+18.9%	0.4195+27.2%	0.4254+28.5%	0.2198+13.2%	0.2241+15.0%	0.1545+13.8%	0.1580+15.2%
B6	0.5401+22.5%	0.5349+18.9%	0.4242+28.0%	0.4254+28.5%	0.2198+13.1%	0.2140+15.8%	0.1546+14.0%	0.1550+15.1%
B8	0.5397+22.7%	0.5373+18.9%	0.4201+27.7%	0.4255+28.5%	0.2130+13.2%	0.2235+14.9%	0.1546+13.8%	0.1567+15.7%
BJI	0.4720	0.4720	0.3706	0.3706	0.2997	0.2997	0.2380	0.2380

Table 1 Statistics on β − 1 or 20, LG = 5

impressed. The fifth principle — a visual has its elements arranged to make its purpose immediately apparent — is certainly not applied here.

To highlight the main point of a visual, its elements need to be organised. In the table in ☞7, the logic chosen for arranging the data does not make the purpose of the visual immediately apparent, even though the arrangement of its elements is not haphazard. The reader has to work hard to isolate the salient data. In this table, the writer actually intended to show that the difference between one-step and two-step methods is small except for the MSV method.

☞7. **Visual gallery of errors: the table without a clear message.** Comparison of all combinations of one- and two-step methods.

Methods	True-positive rate (%)	False-positive rate (%)
BN & BN	22.0	1.3
BN & MO	24.9	1.9
BN & MSV	39.2	0.2
PSY & BN	27.1	2.6
PSY & MO	27.0	2.7
PSY & MSV	66.9	0.3
COR & BN	23.0	1.9
COR & MO	25.8	2.5
COR & MSV	38.1	0.2
BN	21.8	1.2
MO	24.8	1.9
MSV	35.9	0.2

The modified table in ☞8 makes the author's point. However, it still raises questions unanswered in its caption; namely (1) why hasn't the MSV method been combined with the MO method, (2) why are the PSY and COR methods not evaluated as one-step methods, and (3) what do the acronyms mean? (The writer of this table assumes

☛8. **Visual gallery of honours: the clear table.** ☛7 (modified). The comparison of one-step and two-step methods reveals three facts: (1) the improvement resulting from the addition of a second step to the BN and MO methods is minor; (2) the one-step MSV method (35.9% true-positive, 0.2% false-positive) is superior to the one-step BN and MO methods; and (3) adding the PSY as a first step to MSV provides close to a twofold increase in performance (66.9% true-positive).

Methods (1 step & 2 steps)	True-positive rate (%)	False-positive rate (%)
BN	21.8	1.2
BN & BN	22.0	1.3
COR & BN	23.0	1.9
PSY & BN	27.1	2.6
MO	24.8	1.9
BN & MO	24.9	1.9
COR & MO	25.8	2.5
PSY & MO	27.0	2.7
MSV	**35.9**	**0.2**
COR & MSV	38.1	0.2
BN & MSV	39.2	0.2
PSY & MSV	**66.9**	**0.3**

that the reader is familiar with them, but this may not be the case. They could have been explained in a footnote in the caption.)

A graph would also make the purpose of the author clear (☛9).

In summary, be selective in your choice of visual elements. Choose them on the basis of their added value towards your contribution and on the basis of their conciseness (do they make the same point in less elements?). Once chosen, arrange them until their organisation clearly makes your point. Understand that, like the first text version of your paper, the first visual is rarely the best visual. It is only a draft waiting to be improved with the help of readers.

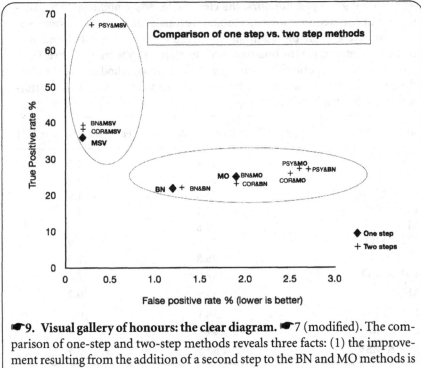

☞9. Visual gallery of honours: the clear diagram. ☞7 (modified). The comparison of one-step and two-step methods reveals three facts: (1) the improvement resulting from the addition of a second step to the BN and MO methods is minor; (2) the one-step MSV method (35.9% true-positive, 0.2% false-positive) is superior to the one-step BN and MO methods; and (3) adding the PSY as a first step to MSV provides close to a twofold increase in performance (66.9% true-positive).

Principle 6: a visual is concise if its clarity declines when a new element is added or removed

Each visual has an optimum conciseness. Visuals ☞8 and ☞9 include the false-positive values. Are these critical to the point the author wants to make? Is it possible to come to the same conclusion without them? Without false positive %, the table is both clear and concise. A footnote could explain why false positives were ignored.

To add visual elements to a graph is so tempting; to merge two graphics into one in order to save space for more text is nearly irresistible. Consequently, the visual often becomes so complex that it is no longer understandable. It becomes the "everything but the kitchen

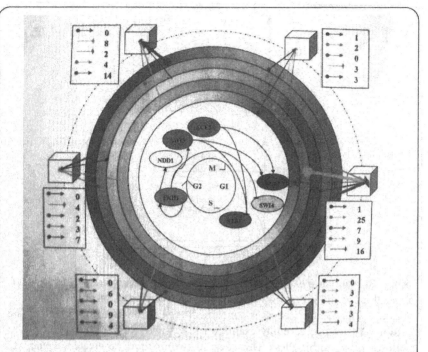

☛10. Visual gallery of errors: the overly complex visual. This beautiful visual combines two related visuals into one. The resulting increase in complexity greatly reduces clarity and understanding.

sink" visual. The density of its elements per square inch hinders rather than helps understanding.

In its original version, the draft diagram in ☛**10** had attractive rainbow colours, 3D elements, arrows, links, and much more. It required much time to design and was a masterpiece. It established a parallel between two phenomena that shared the same cell cycle. It would have been perfect, but for a small problem: only its authors understood it. This figure was later simplified, and clarity was restored.

Even though simplification is often a good remedy, it is not always so. Sometimes, it leads to the loss of key information [☛**11(a)**]. This loss may be felt in other visuals. If adding or subtracting elements in

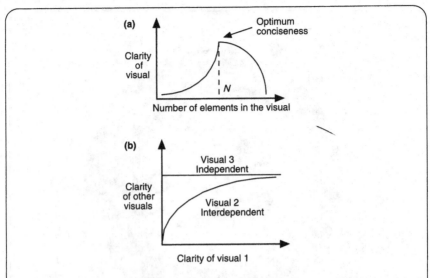

☞11. Relationship between clarity, conciseness, and interdependence. (a) If a visual makes its point with N elements, then any other number will not be optimal as far as conciseness is concerned, even if conciseness is greater. The author will need to carefully examine what is necessary to make that point; nothing more, nothing less. This suggests two methods to create a visual: (1) start with a small number of key elements, and stop adding elements when the point is clear; or (2) start with a large number of elements, and trim until removing elements reduces the understanding of the point made. **(b)** If the clarity of a visual is dependent on the existence of another visual, then the two visuals are interdependent. As a result, the clarity of one visual will affect the clarity of the other.

one visual affects the clarity of another visual, then these two visuals are clearly interdependent [☞11(b)]. They will have to be redesigned in order to increase their independence.

To summarise this point, complexity is born out of (1) a lack of discrimination in the choice of the elements included in a visual, (2) a lack of a clear relationship between the various elements of a visual, and (3) a lack of independence between visuals. Simplification and consolidation increase visual conciseness. But, remember that conciseness is the servant of clarity, not its master.

Examine the complexity of your visual. What makes it complex? Is there a better way to make the same point with less complexity? Visuals come in many types: chart, diagram, table, photo, and list. Would replacing one type with another make your visual clearer? Would dividing one visual into two make things clearer? Would combining two visuals make things clearer? Would reorganising the information in your visual make the relationships between its elements more obvious (using arrows, colours, or words; or sorting the data in a different way)?

Tsunami

In December 2004, a tsunami devastated part of Indonesia. When you recall that event, what comes to your mind? Text? No. Headlines? Maybe. Images? Certainly. It was through images that most of us assessed the extent of the tragedy, not through paragraphs of stirring prose. When we open a newspaper, images (and titles taken as images) capture our attention. We look at them in priority because our brain gathers much information from them in a short amount of time.

That day, when I opened the magazine, I saw a picture of the impact of the tsunami, a picture I will never forget. In that magazine, pictures occupy more space than text does. Each picture is striking and rich in context; it often spreads over two pages. Taken alone, each tells a full story. The caption, subservient to the picture, provides additional details to help the reader answer the where, when, who, what, and why questions. The magazine editors know that their readers tend to flip through pages to learn the facts from visuals. Their readers, first

(Continued)

(Continued)

convinced by the strength of the image content, read the text that falls outside visual and caption last. Are readers of scientific journals different?

Principle 7: besides the caption, a visual requires no external text support to be understood

The strange oasis

An old Bedouin likes to tell the tale of a strange oasis he once came across in the Sahara desert, after a sandstorm had stranded his caravan. The tallest man, who was perched on top of the tallest camel in the caravan, saw it first. "Oasis straight ahead!" he shouted. Everyone's tongue was as dry as paper. There should be water, coconuts, and dates there. They pressed ahead. A short distance away from heat and thirst relief, the travellers noticed that clusters of full coconuts were sitting on the sand dunes, away from the coconut trees in the oasis. Their skin was soft, but they were hot to the touch, so the people took them inside the oasis to drink later. The oasis was small, it had no well, and all the coconut trees were barren, so the only refreshment would have to come from the coconuts found on the sand dunes. Unfortunately, these coconuts were not ordinary. Their husks hardened like steel as soon as they were inside the oasis in the shade. The sharpest dagger could not cut through them. So, the people had to go back out into the desert to open them and then return to the oasis to drink them, a process they found most unpleasant.

The oasis is still there, he claims. It is now an attraction for tourists who go and visit it by helicopter (camel rides are just too slow nowadays). Like the Bedouin, they have to go into the hot sand dunes to get the coconuts.

Nowadays, readers are pressed for time. So, they parachute themselves directly into your paper and have a marked preference for the pleasant visuals, which are far more refreshing than paragraph text. However, they are frustrated because, to understand the visual, they need to refer back to the text and search for "(*see Figure X*)" in the text; then identify the beginning of the sentence or paragraph where the writer explains the visual; and finally go back and forth between the explanatory text, the visual, and its caption until their understanding is complete. This time-consuming and iterative process is most unpleasant.

To accommodate the nonlinear reading behaviour of scientists, each visual should be self-contained or self-explanatory. Its title and lengthened caption should create thirst for the visual and explain it fully. Does the text in the body of the paper remain the same? No, it is shorter and only states the key contribution of the visual, without detail, repeating only what is absolutely necessary to move the reader along. Doing so has two advantages: (1) the visual can be understood without the need to read the whole article, and (2) the body of the paper is shorter (and thus faster to read) because it keeps to the essential.

To illustrate this method, let us examine a visual (title, caption, and photo) and its description in the text. We will rewrite the caption to make the visual self-contained; and we will also rewrite the text description to avoid repeating the caption, and focus instead on the significance of the visual, its *raison d'être*.

In the original visual in ☞12, the CALB and TEM acronyms are undefined, and the readers have to go back and forth several times between text, visual, and caption before they get the full picture. Compare this visual with the modified one in ☞13. The new visual

(Visual)

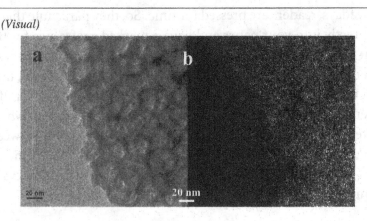

(Caption)

Figure 5. (a) TEM micrograph of CALB/MCF-C$_{18}$ from pressure-driven method, and (b) the corresponding ELS elemental mapping of N.

(Text in body of paper)

"Figure 5 illustrates the uniform nitrogen mapping over the CALB/C$_{18}$-MCF sample, indicating the homogeneous distribution of the nitrogen-containing enzymes within the mesoporous silica matrix. CALB/C$_{18}$-MCF also showed PAFTIR peaks at 1650 cm^{-1} and 3300 cm^{-1} (Figure 3c), which were associated with the amide groups of the enzymes, confirming the enzyme incorporation."

☛12. **A visual that does not stand alone and is not self-contained.** Reprinted with permission from Han Y, Lee SS, and Ying JY, "Pressure-driven enzyme entrapment in siliceous mesocellular foam", *Chem Mater* **18**:643–649, 2006.

is now autonomous, its caption is longer, and the description in the body of the paper is cut down to the essential.

The captions of Figure 5 (and Figure 3 — not shown here) are now self-contained, while the text in the body of the paper is cut to focus only on the point the writer wishes to make (the enzyme is in the porous matrix). Even if the reader goes straight to Figures 3 and 5 and bypasses the text, the same message is given.

(Visual)

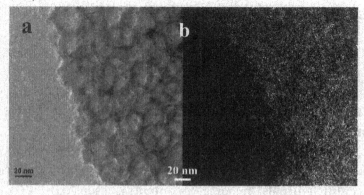

(Modified caption)

Figure 5. (a) Transmission electron micrograph of the *Candida antarctica* lipase B protein (CALB) immobilized by pressure in the porous matrix of hydrophobic mesocellular siliceous foam (MCF-C18); and (b) the corresponding electron energy loss spectroscopy elemental mapping of N. Detection of N is used as evidence that the nitrogen containing CALB enzyme is uniformly incorporated and distributed in the porous matrix.

(Modified text in body of paper)

"Both electron energy loss spectroscopy (Figure 5b) and FT-IR spectrum of CALB/MCF-C$_{18}$ (Figure 3c) confirm the incorporation of the enzyme."

☛13. Visual of Figure 5 now self-contained.

Examine each visual in your paper. Rewrite the caption to make your visual self-contained. Boil down the key contribution of the visual to a sentence or two. Replace whatever you have written about the visual in the body of the text with that sentence.

Purpose and Qualities of Visuals

Purpose of the visual for the reader

1. It allows self-discovery of the paper.
2. It helps readers verify the written claims of the writer.
3. It saves reading time by allowing faster understanding of complex information and faster understanding of problem and solution.
4. It provides a direct (shortcut) and pleasureable (memorable) access to the writer's contribution (in an increasing number of scientific journals, the table of contents is visual).

Purpose of the visual for the writer

1. It makes the paper more concise by replacing many words, particularly in the introduction where it provides fast context, and helps bridge the knowledge gap.
2. It motivates readers to read more, yet allows them not to read all.
3. It provides compelling evidence, in particular evidence of contribution.
4. It enables the writer to represent complex relationships succinctly.
5. It (re)captures the reader's attention and improves memory recall.

Qualities of a visual

A visual is **SELF-CONTAINED**. Besides the caption, no other element is necessary to understand it. The caption and the visual answer all reader questions.

A visual is **CLEAR**. It has a structure, it is readable, and it includes visual cues to help readers focus on key points.

A visual is **CONCISE**. It contains no superfluous detail. It cannot be combined with other visuals without loss of essential information or clarity, nor can it be simplified.

A visual is **RELEVANT**. It is essential to the purpose and the contribution of the writer. It does not distract the reader.

Examine each visual in your paper. Is it concise? Can you hide details in appendices or footnotes? Is the visual essential? Is it understandable to a reader who is not an expert in your field? Is it autonomous and understandable without any support from your main text? Should it appear earlier or later in the paper?

16

Conclusions: The Smile of Your Paper

After ruling out many choices, I decided that the part of the body that best represents the conclusions is the smile. The brain never comes up with such associations completely randomly; therefore, why a smile? I thought again of the many conclusions that had disappointed me and deflated my enthusiasm with self-deprecatory endings such as these: *I have not yet done this; This could be greatly improved; Had I done this, the results would have been much better; For the time being, the performance of this algorithm is still poor; The impact of my research might have been greater if ... instead of....*

I had read these articles with great interest; and right at the end, in the conclusions, I had found suggestions that nothing significant had been accomplished. To explain the extent of my disappointment, I felt like the person who is about to buy a car described as safe, only to discover at the last minute that the car has no air bags and no antilock braking system. Unannounced limitations frequently surface in the conclusions to disappoint the reader who genuinely assumes that the author has already dealt with them. Also, after a convincing demonstration, the author often reneges or casts doubt on his or her achievements. Imagine a lawyer who manages to demonstrate the innocence of his or her client throughout the court proceedings, but who, on the very last day in front of the jury, apologises on the grounds that not enough evidence has been produced to justify the plea of innocence and asks the jury to acquit his or her client with the benefit of the doubt. How unbelieveable! The way a defence lawyer really ends his plea in front of a jury should be adopted to end a scientific paper: with assurance, firmly, and smiling, trusting that the jury will find the client not guilty of scientific insignificance.

You may have noticed that, in some journals, articles have no conclusions headings; the discussion ends the paper. The need to conclude is still there however, even if the heading is absent. Some journals — *Nature* is one of them — recommend to finish an article without conclusions. They would rather have the author write the last paragraph "about the implications of what the reader has read" (*Nature Physics*, "Elements of Style", Editorial, Vol.3, No.9, September 2007), and not summarise what has been accomplished. In some ways, it reinforces the notion that the conclusions are entirely turned toward the reader. Many online tutorials on technical writing, whilst not disagreeing with this view, argue that the conclusions should repeat the main achievements of the author. With good reason, they claim that nothing ever new about the contribution should be presented in the conclusions, simply because it has not been properly backed-up with evidence. I must say that this opinion is hard to disagree with. Following the metaphor of a defense lawyer's final plea in front of a jury, it stands to reason that any attempt to convince the jury at the last minute with evidence that has not been cross-examined is not receivable and is objectionable. Such last-minute-theatrical surprises are the realm of Hollywood movies only. Therefore, "the implications of what the reader has read" should be backed-up by a reminder of the parts of the contribution suggesting or establishing these implications.

It was mentioned in earlier chapters that readers are not always predictable and that they tend to skip large sections of a paper, jumping from abstract to conclusions, like the hurried reporter who only attends the final plea in court. From a writer's perspective, this is not ideal, but you can do nothing to prevent it. If your conclusions differ sufficiently from your abstract, then there is no harm done. Unfortunately, the reader too often finds similar or identical sentences (via "copy and paste") in both. It is therefore necessary to differentiate the conclusions from the abstract to avoid boring the reader. How do the two differ?

1. Often the journal recommends the use of the past tense in the abstract, Unfortunately, the main tense used in the conclusions is also the past tense. You are looking back to what you did. It is all in the past behind you. Only the facts that have been demonstrated without a doubt, the unquestionable scientific facts, are stated in the present tense. The lawyer says "my client is innocent", not "my client has been proven innocent". The present tense in the conclusions reinforces your contribution. If the journal does not impose the use of the past tense in the abstract, it becomes advantageous to write the whole abstract in the present tense because doing so differentiates conclusions from abstract.

2. Because it has to close the loop that is open in the introduction, the conclusions have to be more detailed than an abstract. In the introduction, you describe a world without your contribution. In the conclusions, you show how the world changes because of it. The conclusions bring closure. It closes the door on the past before it opens doors into the future.

3. Whereas the abstract briefly mentions the impact of the contribution, the conclusions dwell on this aspect to energise the reader. In his book *A Ph.D. Is Not Enough*, Professor Feibelman gives his writer's viewpoint:

"The goal of the conclusions section is to leave your reader thinking about how your work affects his own research plans. Good science opens new doors." [a]

4. The abstract adopts a factual, neutral tone. The conclusions keep the reader in a positive state of mind. Unfortunately, quite often, the conclusions of a paper are written last, when the writer's energy is at its lowest point. Think about this when you write your conclusions. Remember that a reader may need to find the motivation to read the rest of your article in your conclusions instead of your introduction. Keep your energy level high.

5. Everything in an abstract is new to the reader. In the conclusions, nothing is. The conclusions do not surprise the reader who has read the rest of your paper. Even the section about future works is expected. In the discussion section, you venture new hypotheses to explain some results, or discover that using different methods might be helpful to avoid undesireable limitations and get better results. The reader who has read your discussion therefore anticipates that, in your future work, you will explore these new hypotheses or use these different methods.

Purpose and Qualities of Conclusions

Purpose of the conclusions for the reader

1. They bring better closure to what has been announced in the introduction by contrasting precontribution with post-contribution. What was unproven, unverified, unexplained, unknown, partial, or limited is now proven, verified, explained, known, complete, or general.
2. They allow readers to understand the contribution better and in greater detail than in the abstract so as to evaluate its usefulness to them.

---

[a] Feibelman PJ, *A Ph.D. Is Not Enough: A Guide to Survival in Science*, New York, Basic Books 1993.

Purpose of the conclusions for the writer

1. They restate the contribution, with a particular emphasis on what it allows others to do.
2. They propose new research directions to prevent duplication of effort or to encourage collaboration.

Examples and counterexamples

In the following example, the author repeats a main aspect of his contribution already announced in the discussion section. It is an encouragement for others to use his method.

> *Our method has been used to determine the best terminal group for one specific metal–molecule junction. We have demonstrated that, in principle, it can be applied to other couplings.*

It is not always necessary to have conclusive results to conclude. Sometimes, the hypothesis presented in the introduction can be only partially validated. The choice of words to say so is yours, but you must admit that the phrasing is quite critical here. Which of these sentences is better?

> *In conclusions, our modified gradient vector flow <u>failed to</u> demonstrate that...*

> OR

> *In conclusions, our modified gradient vector flow <u>has not been able</u> to demonstrate that...*

> OR

> *In conclusions, our modified gradient vector flow <u>has not yet provided definitive evidence for or against</u>...*

The last sentence is much better, isn't it? The word "*yet*" suggests that this situation may not last. Far from being despondent, the scientist is still hopeful. In fact, "*yet*" creates the expectation of good news in the sentences that follow.

In conclusions, our modified gradient vector flow model has not yet provided definitive evidence for or against the use of active contour models in 3D brain image segmentation. However, it **confirms** *that polar coordinates, as suggested by Smith and Al [4],* **are** *better than Cartesian coordinates to represent regions with gaps and thin concave boundaries. In addition, the need for a priori information on the region being modelled is now removed at no cost to the performance of the model.*

The findings are inconclusive, but they reveal that (1) an undesireable constraint has been removed; and (2) for a particularly complex type of contour, another coordinate representation scheme is more efficient. Note the use of the present tense (in bold font) to reinforce the conviction and authority of the author.

Even partial achievements are important to the scientific community when they validate or invalidate other people's theories and observations, and when they establish the benefit of a method against other methods for a particular type of experiment. Science explores, step by step, a labyrinth with many dimensions. Marking a dead end before turning back is sometimes necessary.

In the next example, the findings are conclusive. They could have been even more conclusive, but the researcher wanted to publish them before exploring new possibilities. If the results are promising enough, then why wait until all the possible paths have been explored before submitting a paper. Mentioning what you intend to do next may protect you from competition or may possibly encourage others to collaborate with you.

The 15%–25% improvement in reranking the top 10 documents by using words adjacent to the query keywords found in the top five documents demonstrates the validity of our assumption.

We anticipate that the high-frequency but nonquery key-words found in the top five documents may also improve the reranking, and plan to include such keywords in future research.

Readers may see your limitations as great starting points for their research. Relaxing an assumption or finding a way to bypass a constraining limitation may allow them to make use of your work to solve their problems. In the end, you win because your work has been useful and because you will be cited in their next paper. As you can see, taking the time to state assumptions and limitations is not only good scientific practice, but also a way to promote science and your name in science.

Should limitations reappear in the conclusions, or should they remain in the methodology and discussion sections? If you can present them in a positive manner as future work, then state them again in the conclusions.

Finally, we summarise the limitations of our optimising algo-rithm and offer our future research plan.
• Parameter tweaking. As discussed in section 4.2, the value of alpha is obtained without difficulty, but a satisfactory gamma value is obtained only after experimenting on the data set. <u>We have given the reader pointers to speed up the determination of gamma in this paper.</u> We plan to investigate a heuristic method that allows direct determination of all parameters. In this respect, we believe that Boltzmann simulated annealing will be an effective method.
• ...

Notice (underlined hereabove) the way the parameter tweaking limitation is minimised by emphasising that a method has been given to speed up the labour-intensive part of the algorithm.

When it comes to conclusions, be conservative and exercise restraint. Do not destroy your good work with sentences like these:

> *In the future, we would like to not only validate the clustering results from the promoter binding site analysis, but also incorporate more information such as the protein–protein interactions, pathway integration, etc. in order to have more convincing and accurate results.*

As a reader, what is your impression of the achievements? Do you feel that the author is pleased with his contribution? How about this next sentence:

> *In the future, we intend to experiment our approach using larger data sets.*

Do you think the results are statistically significant? Would you trust the conclusions?

You could use the effective *although* sentence to reinforce your contribution while simultaneously mentioning present or future limitations. Be careful, however.

> *Although these protocols will continue to change, we believe they represent a reliable starting point for those beginning biochip experimentation.*

Despite having the positive contribution in the main clause, the previous example has been negatively perceived by some readers. Why? It may be due to the "*we believe*" statement. Read this sentence again by skipping "*we believe*", and you may find the protocols more appealing. The facts seem to speak for themselves, without the need for beliefs to influence the decision of the reader.

> *Although these protocols will continue to change, they represent a reliable starting point for those beginning biochip experimentation.*

In the next sentence, both main and subordinate clauses contain positive facts. Since the main clause contains information about the future, the future should appear more appealing. But, this is not exactly the case:

> *Although the model is capable of handling important contagious diseases, new rules for more complex vectors of contagion are under construction.*

Even though both the subordinate and main clauses establish positive facts, the overall perception is not always positive. Why? The readers are confused. Ordinarily, if the *although* clause contains a positive argument, the reader expects the main clause to negate or neutralise the value of that argument. In this case, the main clause also contains a positive argument. As a result, the overall impression is mixed.

Before we introduce the qualities that you should build into your conclusions, let us repeat one last time that a paper forms a coherent whole. It tells one story, which is consistent in all its parts. The conclusions are tied to the abstract and the introduction. They support the claims made in both. In addition, avoid establishing coherence through the expedient practice of cutting and pasting.

Qualities of conclusions

Conclusions are **POSITIVELY CHARGED.** They maintain the excitement created in the introduction.

Conclusions have **PREDICTABLE** content. There are no surprises. Everything has been stated or hinted in the other parts of the paper.

> Conclusions are **CONCISE**. Restate the contribution alongside its potential impact. Close the door. Open new doors.

> Conclusions are **COHERENT** with the title, abstract, and introduction. They are a part of one same story.

Examine your conclusions. How positively charged are they? How consistent are they with the claims you made in the abstract and introduction? Do they "open new doors"?

Future Works

My work ends here, and now yours starts. Writing a book is not easy. Sometimes, only after rewriting and rereading a chapter for the nth time does its structure finally appear. Whenever the structure takes shape, the writer feels the joy of the potter seeing the clay change into a vase. At other times, the structure of a chapter is in place even before the contents, and the hard work consists of finding the examples and metaphors to make things clear. But, one thing is constant: the longer you spend rewriting, the clearer your paper becomes.

Writing this book took longer than expected, but then again, I expected too short a time. I was ready to publish after my second draft. Looking back on this episode, I still laugh. How foolhardy of me! I had not considered that once published, the book would be unchangeable, forever clear and engaging or forever obscure and unattractive.

Six months ago, I was not even close to the draft that you are reading today (a publication is nothing but the latest draft of one's writing). I recall the memorable words of Marc H. Raibert, President of Boston Dynamics, ex-head of the Leg Lab at CMU and MIT: "Good writing is bad writing that was rewritten." How true.

Writing is hard. To avoid making it harder than it already is, start writing your paper as soon as you possibly can. It will be less painful and, at times, even quite pleasant. At the beginning, write shorter papers (e.g. letters to journals). You can write more of them, and get a few accepted. Along the way, a few good reviewers will encourage you and pinpoint your shortcomings, while a few good readers will tell you where you lack clarity.

Each chapter in this book contains exercises that involve readers. Value your reader friends. They spend time to read your paper. That time is their gift to you. Accept that gift with a grateful heart, and accept their remarks without reservation. Do not take negative remarks personally; instead, consider them as golden opportunities towards improvement. Do not try to justify yourself because, in the end, **the reader is always right**. Accept readers' questions, and do not think that answering them face to face will help you. Readers of a scientific journal do not have the privilege of having you by their side to explain. Just take note of the remarks and questions, and work to remove whatever has caused your reader friends to stumble. On occasion, thank them for their help. Being French, I recommend giving them a bottle of red Bordeaux for their services, but feel free to offer other vintages or to offer to review their own papers.

Let your introduction convey a research that is exciting, and let your conclusions leave the readers enthused as they look towards the future your research has opened. Show the world that scientific papers can be interesting to read. Create expectations, drive reading forward, sustain attention, and decrease the demands on your readers' memory. To make reading as smooth as silk, iron out the quirks in your drafts with the steam of your efforts.

May the fun of writing be with you.

17

Additional Resources
for the Avid Learner

First, I would like to congratulate you. If you are reading this chapter it is because you have decided to further your understanding of the science of scientific writing by venturing online. I have spent many years exploring the rich humus of the Internet garden and collected about 50 sites worthy of your attention. Their URLs have been tested at the time of publication. They may change over time but you will always find an up-to-date list on my website: www.scientific-writing.com.

Websites on Visuals

Graphics and visuals are essential in scientific papers. Only a few websites deal with the topic. I like the site from Bates University "How to Write A Paper in Scientific Journal Style and Format", and in particular the page entitled "Everything You Wanted to Know About Making Tables and Figures".

http://abacus.bates.edu/~ganderso/biology/resources/writing/HTWtablefigs.html

Naturally, the Grand Master of visualisation is Edward Tufte. On the page "Cancer Survival Rates: Tables, Graphics", I enjoyed the example given that demonstrates that a table is not the poor man's tool, and that it can overtake the usual corresponding graphic in terms of clarity and interest, when the two are blended together.

www.edwardtufte.com/bboard

(search for cancer survival rate on the page)

Websites on Grammar

I have to start this section by mentioning a site you cannot bypass (The French say "incontournable"). Better still, download the PDF file when you find it (or buy the book), print it, and read it once a year. Professor William Strunk, Jr. wrote a little book to encourage his students to write English properly (Grammar, English composition). He intended it to be concise and to the point, which is why it is still read by generations of writers even today. Its title is "The Elements of Style".

www.bartleby.com/141/

www.cs.vu.nl/~jms/doc/elos.pdf

Another article I strongly recommend is the one written in the *American Scientist* by George Gopen and Judith Swan. It touches on aspects of writing rarely covered but essential for fluid reading: text progression.

www.americanscientist.org/issues/feature/the-science-of-scientific-writing

(or go to their home page and search for "the science of scientific writing")

If this article interests you, make sure to read George Gopen's book. "Expectations: Teaching Writing from the Reader's Perspective".

www.amazon.com/Expectations-Teaching-Writing-Readers-Perspective/dp/0205296173

www.worldscibooks.com/general/6286.html

Most websites on scientific writing are found, not unexpectedly, on university websites in academia land. The term "scientific writing" is often synonymous with "technical writing" or "academic writing". One definition of "technical" in the Oxford American Dictionary characterises all three: these writings require "special knowledge to be understood". The differences with non-technical writing extend beyond the knowledge required. They even affect the grammar and the writing style. That is why I favour the grammars written by scientists and engineers. Among those, I recommend "Grammar, Punctuation, and Capitalization: A Handbook for Technical Writers and Editors", written by Mary K. McCaskill from the NASA Langley Research Center in Hampton Virginia (USA). The 108-page grammar book is available in PDF version at the following URL:

www.sti.nasa.gov/publish/sp7084.pdf

Large universities help their students improve their writing skills by making available resources online. I recommend you visit three sites. I particularly like the one hosted by Pennsylvania State University because many online exercises are proposed, and because they constantly refer to Michael Alley's great book, "The Craft of Scientific Writing".

www.amazon.com/Craft-Scientific-Writing-Michael-Alley/dp/0387947663/

The header on top of the web page says "Writing Guidelines for Engineering and Science Students". *Writing*, here, covers an extensive range: from the journal article to proposals, lab reports, theses and dissertations, and even résumés and correspondence. Click on the link "Writing Exercises" under "Student Resources" at the top left of the page, and have fun learning.

www.writing.engr.psu.edu

The Purdue university Online Writing Lab (OWL) is extensive.

http://owl.english.purdue.edu/owl/

The University of Wisconsin has a great handbook containing an extensive grammar section.

http://writing.wisc.edu/Handbook/GramPunct.html

If English is your second language, you probably struggle with the use of the articles "*a*", "*an*" and "*the*". Indeed, the incorrect use of the article will immediately betray the non-English background of a writer. Learn from the experts. Three documents will help you: the Writing Center at Rensselaer Polytechnic Institute has a good page entitled "Article Usage" written by John R. Kohl. It is quite exhaustive.

www.rpi.edu/web/writingcenter/esl.html

Be sure to also read section 1.5.1 in the NASA grammar for a different perspective, and the page entitled "The Use and Non-Use of Articles" in the grammar of Purdue University writing lab.

http://owl.english.purdue.edu/owl/resource/540/01/

Finally, punctuation makes a significant difference in the clarity of writing. Do make sure to read on the use of punctuation. Each of the websites mentioned above has a section on punctuation. The NASA grammar dedicates 32 pages to this important topic, from section 3.1 to 3.16. I also enjoyed reading two pages from the Purdue University Online Writing Lab (OWL) entitled: "Commas with Nonessential Elements" and "Proofreading for Commas".

http://owl.english.purdue.edu/handouts/grammar/g_commaproof.html

http://owl.english.purdue.edu/handouts/grammar/g_commaess.html

Websites on Evolution of Scientific Writing

Alan Gross and Joseph Harmon have written interesting articles. If you subscribe to *The Scientist* online, do read the article entitled "What's right about Scientific Writing" (The Scientist, 134(24):20–, 6 December 1999; you need to be an online subscriber of the journal to be able to read this article)

www.the-scientist.com/article/display/18803/

and then read Michael Seringhaus in a later issue of *The Scientist* (21 November 2006, I think) arguing against the scientific paper in its current form. His article is entitled "The Death of The Scientific Paper".

http://papers.gersteinlab.org/e-print/paperdeath/preprint.pdf

Gross and Harmon have one article freely available online. It is entitled "The Scientific Article: From Galileo's New Science to the Human Genome".

http://fathom.lib.uchicago.edu/2/21701730/

If you want to discover the views of Roald Hoffmann, 1981 Nobel Prize laureate, on scientific writing, and how he wishes it to evolve in the future, read the fascinating article he co-wrote with Pierre Laszlo entitled "The Say of Things" published in the *Social Research Journal*, Vol. 65, No. 3 (Fall 1998).

www.roaldhoffmann.com/pn/modules/Downloads/docs/The_Say_of_Things.pdf

Professor Henrietta Tichy, a specialist in technical and scientific writing, has a few choice words aimed at the old guard of science writers. Her article entitled "Advice To Scientist-Writers: Beware Old 'Fallacies'" (published in *The Scientist* 2(20):17, 31 October 1988) starts with the following sentence: "Bits of advice from fallacy land have a strong influence on writing". You can read the rest — as long as you are an online subscriber to the journal!

www.the-scientist.com/article/display/8862/

Another scientist, cancer researcher Chris McCabe, is on "a mission to sex up scientese" (title of an article published on the *Guardian Unlimited Newspaper* website on 4 February 2004, in its Life section).

www.guardian.co.uk/science/2004/feb/05/research.science

Websites on Persuasion

You will have to convince the journal that your paper deserves its attention. There are two documents I have found that help in that respect. The first is an extensive page written by editor Kwan Choi. The page contains pdf files on topics such as "General publication strategies", "Writing strategies", "Preparation and submission", "Rejection and revision", and even a topic on "being a good referee".

www.roie.org/how.htm

The second document is a paper entitled "How to Write a Frequently Cited Article" by Henk van der Vorst. If you ever want to know how citations work, this article is for you. Do not get caught trying to use your influence to increase the number of citations. On the positive side, there are things the author can do to attract attention in a good way, and the author of the paper reviews them.

www.austms.org.au/Publ/Gazette/2004/May04/vorst.pdf

If your paper is rejected, it is because somewhere, somehow, you did not convince the peer reviewer that it is worth publishing. A good article published in a medical journal reviews the reasons for rejection. Trisha Greenhalgh wrote an article for reviewers entitled "How to Read A Paper: Getting Your Bearings (deciding what the paper is about)". One of its subtitles is "The Science of 'Trashing' Papers". The paper is about clinical research studies.

www.bmj.com/cgi/content/full/315/7102/243

Another interesting paper also in clinical studies is written by Arthur Croft. It is entitled "Scientific Writing: Discriminating Good from Bad". You will find his article at the following URL:

www.chiroweb.com/archives/11/20/01.html

Logical fallacies abound in scientific writing. Scientists unknowingly fall prey to logical fallacies. It matters to be aware of what they are. An excellent article by Stephen Richardson entitled "Logical Fallacies in Scientific Writing" disappeared from its original website, possibly because it applies logic to the heated arguments between creationists and evolutionists, but reappeared here:

www.cs.auckland.ac.nz/~cristian/i2rcs/i2rcs_docs/logic.htm

Websites on Scientific Paper

I enjoyed the Bates College page entitled "The Structure, Format, Content, and Style of a Journal-Style Scientific Paper". It systematically goes through all parts of a paper.

http://abacus.bates.edu/~ganderso/biology/resources/writing/HTWsections.html

http://abacus.bates.edu/~ganderso/biology/resources/writing/HTWtoc.html

I also enjoyed the 20-page guide written in January 2006 by Sarah Deel from Carleton College. She recommends to write the abstract in the past tense — I prefer to follow the guidelines of NASA on this one because the abstract needs to attract the reader into the paper, and the present tense does that better than the past tense, yet I like her style of questions and answers. The title of the guide is "Lab Report Guide: How to Write in the Format of a Scientific Paper".

http://serc.carleton.edu/files/sp/carl_ltc/quantitative_writing/examples/lab_report_guid.pdf

The final document of interest is entitled "Scientific Writing Booklet". It is compiled by Dr Marc Tischler from the University of Arizona. It is a short 24-page booklet that, besides covering the various parts of a paper, explains the differences between active and passive voice and when to use which (see the evolution of writing style sites for more discussion on the use of the passive voice in scientific writing — and in particular the paper by Lilita Rodman).

www.biochem.arizona.edu/marc/Sci-Writing.pdf

www.jacweb.org/Archived_volumes/pdf_files/JAC2_Rodman.pdf

Do read the set of slides prepared by Simon Peyton Jones from Microsoft Research in Cambridge, England. It is entitled "How to write a great research paper". I do not agree with everything he says: his definition of the introduction is too narrow (See chapters 13 and 14 of this book for more details), and I suspect that leaving all related work for the end of the paper is too late. Related works are most welcome in several sections of the paper: in the introduction, the discussion, and sometimes the methodology.

http://research.microsoft.com/en-us/um/people/simonpj/papers/giving-a-talk/writing-a-paper-slides.pdf

I appreciated the brevity and common sense recommendations of two web pages from Professor Railsback from the University of Georgia entitled "Some Thoughts on Writing a Scientific Paper or Thesis" and "Some Comments on Writing and Editing (the latter is as important as the former)".

www.gly.uga.edu/railsback/writing1.html

www.gly.uga.edu/railsback/writing2.html

Finally, There are two documents scientists cannot ignore because each and everyone is responsible for the reputation of Science (let alone our own reputation); The first one, a 76-page PDF document published in 2006 by the Council of Science Editors (CSE), is entitled "CSE's White Paper on Promoting Integrity in Scientific Journal Publications". It has two main parts: 1) "Roles and responsibilities in publishing", and 2) "Identifying research misconduct and guidelines for action". You will find it at:

www.councilscienceeditors.org

The second document comes from my favourite website:

www.the-scientist.com

They have an online daily, which I scan every day. All scientists, not just those in life science, will find something of value in it. The March 2, 2007 issue had an article on a "glossary of retractions", listing out the various degrees of retraction, from the simple correction, to the expression of concern, the partial retraction, and the total retraction with or without permission. For each situation, a sample letter is given.

Websites on Writing Process

Bates College has a web page entitled "A Strategy for Writing Up Research Results" which covers the whole writing process, including the all essential peer-review.

http://abacus.bates.edu/~ganderso/biology/resources/ writing/HTWstrategy.html

The peer review occurs after your draft is complete. You have made sure your logic is sound and you have tried to make your paper as concise as possible without loosing clarity. You then show your article to a reader not familiar with your research. If that reader is used to reviewing other people's papers, then you are all set; more often, however, you may have to explain what you expect of him or her. The peer-review form put together by Bates College is great. It helps a reader, even you, check out a paper methodically before sending it to the unknown super-reviewer whose publishing veto power will mean everything to you and your future.

http://abacus.bates.edu/~ganderso/biology/resources/peerreview.html

The OWL writing centre from Purdue University has a page entitled "General Writing Concerns (Planning/Writing/Revising/Genres)". It has many links to the extensive contents of its site divided into four categories, two of which are about the writing process: 1) "Planning/Starting to Write" and 3) "Revising/Editing/Proofreading".

http://owl.english.purdue.edu/handouts/general/

I recently found a 55-page "Guide to Academic Writing" from the University of the western Cape, completed in September 2003, written by Nelleke Bak. Although its emphasis is on writing a thesis, many of its topics are applicable to the writing of a scientific paper. There is an interesting section on literature review and on critical reading (you will need to read many papers to prepare the related work section of your paper). It also covers the topics of plagiarism and citations.

www.google.com

Search with the following words:

Nelleke Bak Academic Writing Guide September 2003

Her paper comes on top of the list.

Books on Scientific Writing

A PH D is Not Enough

by Peter J. Feibelman

The book is worth buying. I bought twenty copies and use it in my seminars on scientific writing. With 109 small pages of large text divided into eight chapters, the book is read in a few hours only.

I like this book because every time I re-read it, I discover new wisdom. Professor Feibelman is not only an experienced scientist, he is also a good writer. You will find information on his book at the following address:

www.amazon.com/Ph-D-Not-Enough-Survival-Science/dp/0201626632

You may be curious to know how I use this book in the seminar: I have the participants read chapter 4 entitled "Writing Papers: Publishing Without Perishing", 13 short pages in all. I then put participants into groups and ask them questions on the writing process, the relative importance of the various parts of a scientific paper, a successful publishing strategy, and the qualities that a successful publishing scientist should possess. This chapter contains the answers to all. Its richness is astonishing. No wonder Carl Djerassi, in a quote prominently featured on top of the cover, writes "It took me over forty years to learn from experience what can be learned in one hour from this guide".

Scientific Style and Format

by the Council of Science Editors (CSE)

This massive book (658 pages), now in its seventh edition (2006), is an authoritative reference book. You will find it here.

www.councilscienceeditors.org/publications/style.cfm

The book cover subtitle is "the CSE Manual for Authors, Editors, and Publishers". It is a resource that you will probably not read in its entirety. It has 32 chapters divided in four sections. I spent time adding the number of subheadings one to four level deep, and found 935 of them! You can imagine the level of detail involved here. This book covers the special scientific conventions for writing everything scientific, from the electromagnetic spectrum and astronomical objects and time systems to the chemical formulas and genes, chromosomes, and related molecules.

I use it often to answer questions such as "How do I reference a paper to be published, but not published yet?" or "How do I refer to a paper seen on the Internet?" or "How do I refer to non-text information, like a video, an animation…?"

Chapter 27 is about Journal Style and Format. Of interest will be section 27.7 "Journal Articles and Their Parts". But there is also a chapter on Tables, Figures, and indexes (chapter 30), on proof correction (chapter 32), on copyright (chapter 3), and three chapters on grammar (chapters 5, 6, and 7).

I recommend that you ask your library to buy it if they haven't already got a copy. The latest edition has been updated to take into account the rapid developments in life sciences. So if your research is in life sciences, get the latest edition!

"Dos and Don't of Technical Writing"

by Dennis R. Morgan, an article published page 22 to 25 of the August/September 2005 issue of *IEEE Potentials*.

In this editorial, you find the most common errors authors make when writing formulas. A whole section of the paper covers mathematics. The "general writing" section mostly covers punctuation errors, and the "Word usage" section briefly reviews the most common word errors peppering the text of engineering journals. This article is found online through a simple search on its title, but it should normally be purchased on the IEEE website.

http://ieeexplore.ieee.org/xpl/RecentIssue_jsp?punumber=45

Then search for issue 2005, Volume 24, issue 3; the article is the last on the list.

"Citing and referencing guide: Harvard style"

is a 15-page pdf file from the Imperial College in London. As its title indicates, the article advocates the "Name Year" referencing scheme where the name of the author is immediately followed by the year of publication of the document. This guide is less extensive than the massive book of the Council of Science Editors, but if the journal you are targeting for publication requires you to follow the "Name Year" scheme, this article is for you. The citing guide makes clear how to cite, and the referencing guide details how to reference the usual journal or conference papers (electronic or in print), but also more unusual documents such as podcasts, patents, blogs or online lectures.

Successful Scientific Writing

by Janice R. Matthews, John M. Bowen, and Robert W. Matthews

The subtitle of this book indicates its targeted audience: "A step-by-step guide for the biological and medical sciences". You will find it here:

www.amazon.com/Successful-Scientific-Writing-Step-step/dp/0521789621

A thoughtful and thorough 230-page guide that goes down to the type of practical advice you don't expect, as in page 71 of the book in a subheading about writing momentum, and how to keep it.

> "If you must eat [while writing], choose foods that require only one hand, don't stain papers, and don't make you thirsty or (worse yet) sleepy."

Thankfully, computer screens are vertical (for the time being anyway — e-readers are coming soon), but keyboards and journals are not, so watch out for these coffee stains!

This book is thorough and complete. It answers many of the questions the novice writer has when facing the daunting task of writing a first draft — including answers to the dreaded writer's block (pages 72 and 73). A large 24-page chapter on visuals entitled "Supporting the text with tables and figures" is well worth reading.

And naturally, three chapters on grammar and composition complete the book. These chapters are not dry. They are always related to the scientific article, making the grammar points immediately relevant to the reader/writer. The font size is small (10 point) so there is a lot of material to digest in this book (including many tables). The authors break down the text in numerous subheadings. It can be overwhelming at times, and not all information given is of immediate use. So read selectively, but take the time to do the many exercises given (the correct answers are given at the end of the book). Although the guide is targeted for life science researchers, much of it applies across the board to all scientific domains. This is probably the type of book you can read many times and always learn something new as your writing skills improve.

Expectations-Writing from Reader's Perspective

by George D. Gopen

George D. Gopen is a very interesting and provocative writer. I reviewed his book on Amazon.com:

www.amazon.com/Expectations-Teaching-Writing-Readers-Perspective/dp/0205296173

Here is the review:

"I have been teaching Scientific writing. I helped many scientists but nobody was helping me understand why the sentences I had rewritten were far better than the originals, why paragraphs had regained energy and clarity. This made it difficult for me to transfer my knowledge to others. Then … through a *Scientific American* article, I discovered George Gopel and this book. I appreciated George Gopel's thoughtfulness. Part two of his book is for us, teachers of writing. Entitled "Pedagogy", it shares George's methods; It also answers the students' common objections to the reader approach. The book's appendix provides an interesting historical perspective on the topic. I recommend this book without reserve."

After reading this book, while home schooling my son, I discovered to my great surprise that French schoolteachers teach 9th graders the same grammatical principles. Who likes grammar! Is it not enough to enjoy reading and not worry about the reason for our enjoyment? Text progression techniques are essential to reading speed and clarity of understanding, yet they are too succinctly presented in books on writing, often under the title "transitions". What a shame!

Craft of Scientific Writing

by Michael Alley

I already mentioned the Pennsylvania State University website which provides grammatical exercises. Many of these advise the reader to look for answers in chapters of Michael Alley's 1995 book. You will find more information on the book on the Amazon webpage, including some eloquent reviews of the book.

www.amazon.com/Craft-Scientific-Writing-Michael-Alley/dp/0387947663

When I was asked to conduct scientific writing seminars, at the turn of the last century — it's not that long ago — this was the first book I selected to prepare the course. Being French (we don't like rules, but we love principles), and having discovered that participants sent to the class felt as if they had been sent to the gulag, I decided to move away from the dryness of grammar. Michael's book was a refreshing departure from grammar and sentence structure books. Like the book of Professor Feibelman, the pages in this book are small and the large text readable. The grammatical part is pushed back to the appendix "avoiding the pitfalls of grammar and punctuation". The examples are not too technical and the author frequently mentions the readers, thereby balancing reader and writer points of views. The book gave me the idea to change the focus of attention from the writer to the reader. All participants to the writing class are expert readers. They can relate better to the solutions proposed to improve their writing after they understand what writer-created problems cause inefficient reading. Later on, the book by George Gopen allowed me to refine my understanding of the reader and I was able to improve on Michael Alley's chapter 9 ("Language: Being Fluid").

The book focuses on structure (chapters 2 and 3), language (chapters 3 to 9), and illustrations (chapter 10 and 11). These chapters help you identify the difference between good and bad writing. The remaining chapters are only brief introductory chapters on presentations, writing genres, and the writing process.

Writing for Computer Science

by Justin Zobel

Here is the Amazon.com page for all information on this excellent book, as well as reader reviews.

www.amazon.com/Writing-Computer-Science-Justin-Zobel/dp/1852338024

My roots are in computer science where I started my professional life. Therefore I was attracted by this title. I borrowed the book from the library and read it from cover to cover. I found myself wanting to use a yellow highlighter pen on the text of quite a few chapters. The pertinence of the examples chosen proves that Justin has reviewed his fair share of papers. If you referee papers, chapter 12 on "Refereeing" is for you. If you are in computer science or if you use computers in your research, then chapter 5 ("Mathematics"), 7 ("Algorithms"), and 11 ("Experimentation") are written specifically for you.

Scientists outside of computer science are not left out. Chapters 2 to 4 are about style and grammar. His examples will bring a smile to your face as you recognise your own mistakes. I cannot say I always agree with Justin's examples. For example page 32, he writes "Beginning a paper by stating that a topic is popular or that a problem is important is flat and uninspiring." I could not agree more, of course. He then illustrates such a flat start with the great example "Use of digital libraries is increasingly common". But then the "may well be preferable" counterexample that follows has the same problem: "Digital libraries provide fast access to large numbers of documents". It uses two imprecise adjectives and does not enhance the knowledge of even the most junior researcher in computer science (see chapters 13 and 14 of this book for more advice). It is easy to critic anyone's work, and I am sure mine deserves its fair share of criticism; Justin's book has the merit of systematically illustrating the principles of writing he recommends to the readers. Chapter 6 on graphs, figures, and tables gives many examples. Justin believes in making figures "less dependent on the paper's text" (page 112) by making their caption more informative. I do too, and even go one step beyond his recommendation: make your figures, tables, and graphs stand-alone through a complete caption. Chapter 9 gives general instructions for writing the various parts of a scientific paper, from its title to its conclusions.

Index

Reader behaviour

Acceptation of ambiguity or incomplete understanding, 6, 7, 47, 94

Direct access inside the paper via the structure — headings and subheadings, 4, 21, 139

Direct access to the visuals, 4, 195

Expectations of new knowledge based on old knowledge, 61

Expectations of elaboration and development, 19, 24, 57, 66, 145, 156

Expectations of justification, 53, 58

Expectation of logical sequential progression, 59

Desire to find within a visual an answer to the questions raised by it, 178, 179, 181

Active reading, constantly anticipating what comes next, 57, 82

Nonlinear reading, 4, 195, 202

Generalisation of doubt on one fact to doubt on whole paper, 152

Rapid discovery of a paper's contents through the headings/subheadings, 4, 136

Likely meaning considered as intended meaning, 6, 7

Pronounced taste for visual information, 29, 177

Frequent reliance on author to understand significance of work, 21, 22, 33

Author behaviour

Lengthening of a sentence by addition of details, 14, 20, 107

Use of synonyms to avoid repetition, 10, 124, 136

Systematic use of the passive voice, 152–154

Sketchy and lifeless introductions and conclusions, 142, 155, 162, 165, 200, 202

Related work section based on reading abstracts or titles only, 172

Research impact left to the reader's appreciation, 33, 121

Long paragraphs, 21

First draft with overlong passages, repetitions, and discontinuities, 68, 155–157

Abstract or conclusion written via "cut and paste" of sentences from the rest of the paper, 23, 158, 202

Abstract containing introductory material, 126

Background all parked in one section of the paper, 11, 12

Paraphrases, 19

Separation of sentences which should be adjacent, 20, 71, 72

Separation of a visual display from its explanation, 15, 195

Visuals not revised once created, 189

Use of acronyms, pronouns, and prepositions without due

Interesting

Organised